The Second Coming
of Science

To David —
Love, Peace + joy !

12-8-92

The Second Coming of Science

An Intimate Report on the New Science

Brian O'Leary

North Atlantic Books, Berkeley, California

Published by   North Atlantic Books
                2800 Woolsey Street
                Berkeley, CA 94705

Cover Design by Paula Morrison
Book Design by Daniel Drasin
Printed in the United States of America

The Second Coming of Science: An Intimate Report on the New Science is sponsored by the Society for the Study of Native Arts and Sciences, a nonprofit educational corporation whose goals are to develop an ecological and crosscultural perspective linking various scientific, social and artistic fields; to nurture a holistic view of arts, sciences, humanities and healing; and to publish and distribute literature on the relationship of body, mind and nature.

*To Meredith, whose beauty and artistic inspiration
have filled my life with golden heart nectar.*

Acknowledgments

I would like to offer my most heartfelt thanks to my friends Paul Von Ward, Maurice Albertson, Bob Siblerud, Gary Zukav, Lawrence and Sylvia Schechter, Roger and Lydia Weiss, Laurene Johnson, Ann Harcus, Lorraine Katena, and to many others, for their invaluable support and encouragement. I want also to express my great appreciation to Daniel Drasin and John White for their perceptive feedback and skillful editing. Most of all I want to thank Meredith Miller for coming into my life as the perfect reflection of my developing openness, sensitivity and right-brained understanding, without which I could not have completed this book.

Contents

Foreword

At the present time, books of this kind may not be for everyone; but I believe that before too long they will be much more widely accepted. I myself wouldn't have picked up this book during the 1960s or 1970s when I was a university-based planetary scientist. At that time I was a typical "left-brained" academic—rationalistic, reductionistic, deterministic, materialistic—and I felt I had the universe and its laws mastered. Serving on the physics faculty at Princeton University I lived a comfortable life and was well accepted by my peers. Yet for all that, I felt alienated. Cut off from life. I felt no joy about the science I was doing.

Then, quite unexpectedly, something happened.

In the spring of 1979, during a weekend "human potential" workshop, I temporarily let go of my rigid thinking. As a result, my hard-earned scientific belief system was irreversibly shaken. In one of the final processes of the workshop, sitting opposite a total stranger who gave me only the name of a man, his age and his town of residence (I knew no one from this town), I was asked to tune in psychically on this man. While the scientist-skeptic within me chattered "poppycock!" I found myself somehow slipping into a state that could best be termed "trancelik," and immediately conjured up an image of this person. Then I described him as a meteorologist-journalist who liked to hang out on the west coast of Maui and who had lost his wife by death.

All of that "poppycock" turned out to be true. Lacking the conventional sensory and informational means of arriving at such an accurate description, I was dumbfounded. My actions appeared to violate the precepts of Western science—*my* science—which had largely formed my sense of reality and security. Without knowing it, I had just participated in a successful "remote viewing" experience—a kind of clairvoyant process known through the ages by psychics, mystics and, lately, parapsychologists.

The next thing I did was to temporarily deny the experience. After all, I had no model into which to fit this "new reality," no box into which to pigeonhole it and, I assumed, no familiar means with which to investigate it. In denial I was able to return with impunity to my comfort-zone, and to my scientific flock.

But not for long. Other workshops and healing experiences, as well as a near-death experience, propelled me along a path of anomalous adventure that time and again transcended the limits of my scientific "truth" and led eventually to the writing of my recent book, *Exploring Inner and Outer Space*. The fact that many of these experiences are corroborated by others and confirmed by rigorous scientific experiments has led me to the conclusion that science itself will need to be revised to embrace these greater truths. That is what this new book is about.

Preface

One day in 1977, a bright senior honors student walked into the office of Robert George Jahn, then Dean of the School of Engineering and Applied Sciences at Princeton University. The student complained that she couldn't find a faculty sponsor for her project, which was on psychokinesis, or "mind over matter." She wanted to build and test an electronic device called a "random event generator," whose output might be influenced by the thoughts of an experimental subject. One might imagine a conversation like this:

"Dr. Jahn, could you sponsor this project?"

"No, not at Princeton. We don't do these things here."

"But you said we could do our project on anything we wanted as long as the experiment is feasible. This one is. And after all, I'm a tuition-paying student."

"O.K., I'll oversee your project. But we'll need to stay quiet about it...."

As we shall see in the following pages, the rest of this story became history. Soon Jahn himself was quietly and cautiously performing experiments on psychokinesis, remote viewing and precognition (predicting future events). Although Jahn and I were acquainted, neither of us then knew of the other's work in the world of the paranormal; so concerned were we both about what our peers would think, that to have shared such information would have been out of the question. But our transformations were similar; we had both discovered the presence of a mind-over-matter "margin of reality," as Jahn has termed it.

We also found that there are many ways in which almost anyone can learn to demonstrate this margin of reality; for example, spoon-bending (something I teach in workshops with virtually 100% success for everyone who tries), firewalking, or something more practical such as healing ourselves and others by visualizing or intentionally sending energy and love. We can even measure the expanded energy field around the one being healed by means of dowsing (what many Russian scientists call "biolocation"). We can also use dowsing to detect which of several glasses of water has been "blessed," or given conscious energy—a distinction we can also learn to detect by taste.

The book you are about to read describes and elaborates on these and many other surprising discoveries with which science, given the philosophical baggage it has carried for over 300 years, will have to reckon as we enter a new century and a new millennium.

A Reverie in 2020 Hindsight

Kona, Hawaii, January 27, 2020

The hazy fog had finally cleared. The sky was a deep blue as I looked up through the skylight of Roger and Lydia's guest house. White cottonball clouds passed rapidly by. The giant ferns of the rainforest were waving in the wind and the smoky eruptions of Kilauea volcano on the other side of the island had subsided at last. It was another beautiful day in Hawaii. That day, I turned eighty.

My body was tired, and sometimes it felt as if my life was gradually waning. I knew I was getting closer to the inevitable transition to the beyond; but I was confident about making the journey because our culture no longer denied the reality of our survival of death. In fact, we now celebrated it as an integral part of our total being.

That hadn't been so in the turbulent twentieth century. Even as recently as the early 1990s it wasn't fashionable to address such cosmic issues through science, even though they had always been susceptible to scientific inquiry. Postmortem survival and scores of other vital questions had been ignored by mainstream science, relegated to the realm of religion, scorned, denied, and disdained.

As a younger man, maybe forty years ago, I had paused one day to ponder these taboo questions that were now either answered or the subject of intense and widespread research. "Do the UFOs and the crop circle phenomenon indicate the presence of alien intelligence? Can our minds affect matter directly? Can we heal ourselves through visualization and laying-on of hands? Can we heal the Earth? Are there dimensions beyond time and space? Is

there a higher organizing power governing all of nature?..."
Science now says "maybe" or "yes," and forms working
hypotheses for studying such questions more deeply. In the
old days, although scientists and educators always claimed to
be above superstition, double standards, and limits on academic freedom, it was sternly forbidden even to ask such
questions. Today's students view this as most curious, and
often as the basis of much creative humor!

I so clearly remembered my struggle during the 1980s
about all that—how I, as a faculty member of the Department of Physics at Princeton University, had started to move
outside the reductionistic science in which I was trained. In
1979 I began to have experiences that appeared to violate the
"rules" Western science had superimposed upon the structure of reality. A psychic experience, a near-death experience and a paranormal healing experience had crumbled
my world view; those things weren't supposed to happen,
yet they did! I couldn't deny them or rationalize them. I had
to explain them—I *had* to.

Prior to that I had learned always to describe reality
as a collection of material things whose characteristics could
be understood only when they were dismantled and reduced
to their component parts. That compulsory dis-integration
was one of the central principles of reductionism, which
essentially attempted to explain the nature of the unknown
in terms of the known, the subtle in terms of the gross, and
the living in terms of the inanimate.

Whether we scientists were looking at atoms,
molecules, cells, butterflies, people, planets, solar systems
or whole galaxies, we had already fragmented them in our
minds, isolated the parts, and seen these immensely subtle
systems as nothing more than machines. Machines whose
components move and interact along limited spatial dimensions over finite periods of time, exclusively obeying
immutable, simplistic principles of physics such as Newton's
Laws.

But in confronting "unexplainable" phenomena, what I learned was that we—scientists and lay people alike—had isolated ourselves in an invisible box of our own making, circumscribed in time by our births and deaths, and in space by a damaged Earth beneath us and the vast realms of seemingly inaccessible space and emptiness above us. We had painted ourselves into a conceptual corner that kept us from asking the most basic questions about ourselves and the universe. As a result we had become prisoners of our own limiting beliefs. Science itself had become a religion, with mechanistic materialism as the Supreme Dogma. This realization came to me as a profound shock.

At the same time I had to acknowledge that the scientific method itself was indeed very powerful. After all, hadn't it helped create the many useful technologies we then took for granted?

What was obvious now (in, forgive the pun, 2020 hindsight) is that we simply needed to separate the scientific method itself from the package of beliefs that surrounded it, and from which it had become almost indistinguishable. Having done so, we were now free to use this powerful tool to address basic cosmic, as well as mundane, questions. As a result we found more meaning in life, and are now well along in solving our most pressing global problems. By using the scientific method without being bound by a narrow philosophy of science, we have, in more ways than one, entered a new millennium.

"No problem can be solved from the same consciousness that created it," Albert Einstein had once said. During the 1980s I concluded that we needed an entirely new consciousness to heal our estrangement from ourselves and from our fellow humans, animals, plants and the Earth itself. We also needed "clean" technologies and we needed them soon. What we needed was a New Science.

That morning in Hawaii, as I looked into the mirror to shave, I reflected on how far we had come, yet how far we still had to go. My ride down to the Kona Hilton was a little late, giving me the time I needed to integrate some of the past into the day's main event.

I was a bit nervous because I felt many younger people didn't know or appreciate how difficult it had been for us in the late twentieth century. But I also knew I had to let go of my "old-timer" image. They had their own lives to live, their own aspirations and challenges, and they were inheriting this world. What I could offer was some wisdom, some passion and some motivation. Most of the work on this evolutionary jump was done. But not all!

As we headed down the hill I thought of the historical significance of the day, my eightieth birthday. It was the thirtieth anniversary of the International Association for New Science, and I was to be the keynote speaker at the Association's gala anniversary meeting. It was a time to celebrate and renew, and to create even bolder concepts—ones of which even I sometimes confess to being skeptical! My younger colleagues will be looking at these as I get closer to my moment of transition. Life must go on.

I dug through the notes for my speech, making sure I still had the quote Carl Jung made while in his early eighties, shortly before his death. He was a new scientist before his time. "I have failed," Jung had said, "in my foremost task to open people's eyes to the fact that man has a soul, that there is a buried treasure in the field, and that our religion and philosophy are in a lamentable state." We have come a very long way since then. But the journey has only begun and I still see some resistance to newer thoughts. Jung had every right to be bitter, I mused.

Then it suddenly occurred to me that I was celebrating yet another anniversary. About fifty years ago, in this very same hotel, I had given a paper at one of the early annual meetings of the American Astronomical Society's Division for Planetary Sciences. I had been a youthful thirty, eager to report on some observations I had made at Cerro Tololo Observatory in Chile, of Mars and Venus. We had just begun the exciting exploration of these new worlds and I felt I was in on the ground floor of it all. Little did I know then that I would be back in the same spot, half a century later, still as a scientist but a very different one than in 1970.

During the 1970s and 1980s the mysteries of the planets and of nature in general unfurled in a relatively methodical, anticipated way. They were interesting, but our mental blinders screened out the greater portion of reality. We were ignoring where the real action was: unorthodox healing, psychic phenomena, UFOs, the crop circles, the anomalous formations on Mars.... Since we refused to acknolwedge their possible reality, we had failed completely to bring our beloved scientific method to bear on them. Our consciousness, our "inner" space, was radically limiting our understanding of "outer" space.

But toward the turn of the century, science came in for an unprecedented cleansing. I shuffled through my notes to find the quote from Thomas Kuhn's classic book *The Structure of Scientific Revolutions.* "The reception of a new paradigm," he wrote, "often necessitates redefinition of the corresponding science. Some old problems may be relegated to another science or declared entirely 'unscientific.' *Others that were previously non-existent or trivial may, with a new paradigm, become the very archetypes of significant scientific achievement.*" I had underlined that sentence myself because it described so well the process that swept us into a new millennium and a new science.

As I prepared to give my talk to eager young men and women of the 21st-century scientific community, my memory wandered to those key experiments of the 1980s and 1990s that had cracked open our cosmic eggs with data that could not be ignored, denied, explained away or suppressed. I would never have dreamed back then how far all this would take us as a culture, as a member of the galactic community.

We had ended up doing what no futurist of the early 1990s dreamed possible. Science, unfettered by the mechanistic-materialistic philosophy and the narrow dogmatism of certain scientists, had supported and nurtured a glorious blossoming of the human spirit. Civilization had been rapidly transformed, environmental disasters-in-the-making had been reversed, and the world was now poised on the edge of the first true Golden Age.

As I came out of my reverie I offered a heartfelt prayer of thanks to the Creator-Spirit of the cosmos, as I felt palpably, and saw in my mind's eye, the mighty intelligence that guides our existence through the long sweep of evolution to higher states of being. It was awesome to contemplate, and I didn't—couldn't—fight the tears of joyful wonder that trickled down my cheek. Could I convey even a small part of this to my waiting audience?

As we pulled up to the hotel driveway I gave my notes one last glance. I walked out of the car toward the conference hall.

2

The Current State of Science

Etymologically, the word *science* simply means "knowing." The primary dictionary definition of science is "knowledge reduced to a system."

But there are many knowledge systems, and each one is also a belief system. Each tends to be quite finite, with well-defined, though somewhat arbitrary, boundaries. Often these boundaries are determined not by the knowledge system itself as much as by the beliefs and disbeliefs of the cultural, historical, religious and political contexts from which that system arose and in which it is embedded.

Western science as we have known it is only one among many possible ways in which people describe what they consider to be their consensus reality. Unfortunately it is unequal to the task of describing our entire experience of what is real, nor does it guarantee that wisdom will arise from the knowledge so described.

The modern Western world believes, however, that its science does define the truth, the whole truth and nothing but the truth. In my own training as a scientist I learned to solve problems by using the scientific method, and because this revealed powerful realities to me I learned to revere its principles and findings. My colleagues agreed with this process and its results, and so their reinforcement and rewards (promotion, grants, emotional support) made it all the more powerful.

What, exactly, is the Western scientific method? As commonly understood, it begins with the collection of observations and data about a particular subject—theoretically, any subject. This is followed by the development of hypotheses that might explain or illuminate the initial observations. Experiments are then devised to test one or more of these hypotheses in a manner that is well-controlled and repeatable by others. The observations, hypotheses, testing procedures and results are then presented in research papers which are submitted to journals and peer-reviewed by experts in the field. Critiques are then incorporated into the next round of experimentation, and the results eventually refined.

A Marriage Made in Heaven?

If the knowledge gained by this method has practical application in the material world, commercially profitable goods and services might eventually spin off into the marketplace. In this way, technology, or "industrial arts," reinforces the scientific view of truth—so strongly, in fact, that technology is often popularly mistaken for science itself!

Taken together, science and technology have set the pace of our society just as the Roman Catholic Church did in Europe during the Renaissance. Science and technology lie at the center of our leading educational, governmental and corporate institutions, and of our economic infrastructure. Automobiles, television sets, VCRs, nuclear power plants and weapons, smart bombs, computers, jet aircraft, rockets, electrocardiograms, medicines, and modern household conveniences are the material backbone of our society. The marriage of science and technology gives us well-promoted benefits.

It has also accrued staggering costs, and the bills are now coming due.

What at first promised to be cheap, convenient and universally beneficial uses of new knowledge are now often seen as serious threats to our physical and psychological health, to the continued viability of the environment, and, through weapons of mass destruction, to the very existence of life on our planet. Sadly, our leading institutions harbor strongly vested interests in promoting these conventional technologies, and by extension this precarious state of affairs. It is imperative, therefore, that we look at this marriage of science and technology more closely. How well has it served humanity? Has it gone past the point of diminishing returns? What went wrong?

I will leave to others the task of detailing the bankruptcy and perilousness of our situation. What I do propose to show, however, is that real solutions can come— not from tinkering with incremental changes to the system, but only from a new and enlarged sense of reality. This new reality may then be reflected and illuminated by a New Science that retains the best and most sensible fundamental principles of the old, while transcending the arbitrary limitations imposed on it by the beliefs that form our current consensus reality.

The Box Metaphor

I am proposing that in current science we have put ourselves into a box. Outside the box lies forbidden territory in which the simple act of asking questions about certain phenomena is essentially prohibited. The fact that many of these phenomena are repeatably demonstrable and easily subjected to observation, hypothesis and rigorous scientific inquiry, makes no difference.

The Canadian theologian John Rossner defined the limitations imposed on us by most of our leading scientists in terms of a box confining us collectively in time and space (Figure 1). On the time axis, our accepted science focuses on recorded history as the legitimate basis of defining reality, with only mundane conjecture permissible about pre-history or the future. Not admissible would be the transcendent or extramundane aspects of reality apparently understood by the "primitive" cultures that produced the pyramids, Stonehenge, the statues of Easter Island, the fortresses of Peru, etc. As for the future, prophecy and pre-cognition are considered illusory in spite of experimental evidence to the contrary. Our very perception and beliefs about time itself are conditioned and constrained by the psychological measuring-stick of the human lifetime and the arbitrary structure of our Western calendrical system.

In the spatial dimensions we are confined above and below. Above us we are bounded by the top of our atmo-sphere, prisoners of our own gravity. The space program, only a generation old, has provided some leakage outside the box; but even these explorations have been made mun-dane by the mainstream's avoidance of such provocative subjects as extraterrestrial life and astrology, despite repeat-ed evidence of various classes of anomalous earth-space interactions.

Below us we are bounded by the Earth. To most Western scientists, Earth means an inanimate geophysical entity providing a foundation for "nature," which is itself viewed as something to manipulate and exploit. But beyond the bottom of the box lies Gaia, a living, self-organizing, interactive entity considered sacred by every culture that has ever derived its sustenance from her bounty. Every cul-ture, that is, but ours.

Could it be that the solutions to our collective, glob-al problems lie outside the box?

Now envision the box that confines the individual (Figure 2). The time boundaries would be birth and death, with questions of pre-existence, near-death experience, sur-vival of death, reincarnation, etc., falling outside the box.

The spatial boundaries would be defined by the limit-ations of the purely physical body. At the bottom they would exclude our own inner experience of physical and emotion-al healing, altered states, psychic experiences, mind-over-matter, etc. Above the top, we would have out-of-body experiences, relationships to nonphysical or spiritual entities, and certain types of UFO-related experiences.

According to the box model, our scientists act as the high priests of our culture, zealously guarding the boundaries. Like billiard balls making endless cushion shots against the inner walls of the box, we seem to follow, in our daily lives, the mechanistic, deterministic assumptions of our present-day science. With research and funding allocations entirely limited to subject matter within the box, the buck stops at the boundaries.

Meanwhile, leakage outside the box becomes possible only in secrecy or at the risk of ridicule and threats to one's career. As a culture we are so conditioned to stay within the box that individual or collective excursions outside it become anomalous events from which we must immediately return, lest we lose credibility or respect from our peers and loved ones.

Going Outside the Box

As we will see, much of what lies outside of our culturally and individually self-imposed boundaries is not only quite open to our experience and demonstration, but is also susceptible to scientific inquiry with the same rigor as that practiced in "authorized" fields by mainstream scientists.

The New Science seeks, therefore, to examine phenomena currently outside the box—to push outward or remove the boundaries that have so limited us. To the New Science, these phenomena no longer are freakish; they are integral parts of our reality.

But how do we take the first steps? One way is illustrated by the famous "nine dot" puzzle. The task is to draw four straight lines through all nine dots without lifting your pencil off the page. Can you do it?

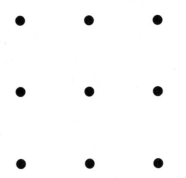

In the end there is only one way to solve the problem: go outside the box!

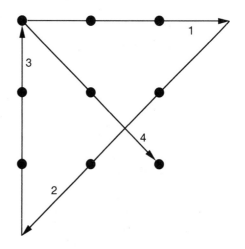

But how do we do that? What is the key?
The key is to be *willing* to go outside the box.

In the real world, an excellent example of "going outside the box" is the ongoing independent research into the "face" on Mars (Figure 3). This two-kilometer-long mesa photographed by the Viking orbiter in 1976 was at first dismissed as a "trick of light and shadow" by NASA-funded scientists who first viewed the images as they came in. The face was pronounced an illusion, not on the basis of any analysis, but on the personal opinion of a few staff scientists. Incredibly, that anecdotal evaluation became NASA's official, "scientific" conclusion.

But a few brave, independent researchers dared to be willing to go outside the box, and raised questions. As a result, other, corroborative images of the face were found in NASA's own archives. Following this, images of other anomalous structures were discovered, some of which seemed to be related geometrically to the face. Later analyses using conservative digital algorithms showed that the three-dimensional structure of the face was indeed facelike and not a two-dimensional trick of light and shadow. Fractal analyses further reinforced the hypothesis that these structures may not be entirely of natural origin. Subsequent studies have suggested that sophisticated mathematical relationships may exist among the anomalous objects.

However, all attempts to publish these results for peer review have faced rejection for reasons related not to the methodology of the research but rather to the field of inquiry. (Since a humanoid face could not possibly appear on a "dead" planet, no legitimate investigation of its possiblity is permissible!) My own attempts to publish one such paper met with several rejections until publication four years later in the *Journal of the British Interplanetary Society*, substantially unchanged from the original version.

New scientists will need to be persistent.

New scientists will also have to be patient, at times, in their development of appropriate standards of proof and repeatability. For example, certain lines of inquiry and experimentation may at first result primarily in descriptive, qualitative or correlative information rather than in fully-formed hypotheses. One case in point might be the type of experiment in which a donor's white blood cells react measurably to stimuli given to the donor, who is now at a distant location. While as yet we have no theoretical foundation upon which to explain this phenomenon, we can repeatably and rigorously document its existence by time-correlating the stimuli and the responses.

In this context, the skeptical argument that "extraordinary claims require extraordinary proof" appears to be increasingly bankrupt. Since yesterday's heresy is often tomorrow's truth, the line between "ordinary" and "extraordinary" will always be essentially subjective and is ultimately impossible to pin down. To act on such arbitrary distinctions is in many instances tantamount to promoting a scientific double standard.

It is also curious that the "extraordinary proof" argument is most frequently invoked as an argument against initiating the very investigative process that might develop such proof. The zealous skeptic often seems to overlook the fact that proof is the final stage of a successful scientific investigation, not its starting point!

I believe we are well beyond the need for this kind of preclusive "rigor." Instead, we need to allow much more work (and funding) to be put into the New Science, using a wide range of exploratory and experimental approaches. As in every field of science, some will succeed more than others, and still others may fail. But, as in every field of science, we will need to try. Once the New Science achieves a certain level of legitimacy and funding we would expect the best scientific minds to be attracted to the field because of its enormous potential. This, in turn, will ensure the maintenance of the highest standards of research.

History as a Teacher

Benjamin Franklin's famous kite-and-key demonstration, magnets pulling along iron filings, the generation of static electricity by rubbing glass on fur and so forth were all "anomalous" precursors to Maxwell's formulation of his equations, and eventually to the development of miracle devices that lit up the night, relieved our physical labors and allowed us to whisper across continents. Our previous "box," bounded by the rules of purely tangible, mechanical forces (as reflected in clocks and steam engines) gave way to a new, larger, more subtle and wondrous universe of electricity and electromagnetics—a universe once inconceivable even to the most sophisticated and imaginative mechanical engineer.

In some respects our position is similar to that of the late eighteenth-century pioneers in electricity and magnetism, and also to that of the early 20th-century relativists and quantum physicists who had to reconcile the otherworldly properties of the very large and the very small with the nature of ordinary, human-scale reality. But I believe that the New Science of today must take even a more fundamental "quantum" leap. As the experiments of Robert Jahn and others conclusively show, we are dealing with the direct interaction of the human mind not only with subatomic particles but with the gross, material world. This demands the development of new paradigms in physics, biology and medicine, to say nothing of new models of consciousness itself.

The interaction of all things, living and non-living, at all levels of complexity, and the bafflingly intelligent tendency of complex systems to self-organize, lie at the heart of the New Science—what futurist Willis Harman calls a "wholeness" science. No longer are we exclusively concerned with the separatist viewpoint that places the observer outside of the observed and in which every measurement seeks to describe the observed solely in terms of its constituent parts. Rather, we are also concerned with functional wholeness whose properties are often transformed by unavoidably participatory acts of observation.

"Great spirits," said Einstein, "have always encountered violent opposition from mediocre minds." The same may be said for each of us; we all suffer to some degree from allowing mediocrity to block our potential. History is replete with examples of resistance to fundamental change when a culture is confronted with anomalies. Most of Galileo's contemporaries refused to look through his telescope because they didn't believe—or didn't want to believe—it could make things look bigger. During the 1700s the French Academy of Sciences refused to believe meteorites existed because rocks couldn't fall out of the sky. And, in 1967, not one of 300 Ph.D. astronomers gathered at a reception bothered to step outside to have a look when it was announced soberly that there was a UFO in the sky outside the building.

It is part of human nature to hang on to older paradigms out of a sense of safety for the ego; funding, power and prestige generally go to those who work within the boundaries of accepted science. This was certainly true in my case until I could no longer function honestly within a framework that did not permit the inclusion of anomalous data.

The Importance of the New Science

The Earth is in desperate need of healing, and the old ways are not equal to the task.

But if we can heal our bodies with our individual minds, can we heal the planet with our collective minds? I think so. Could it be more than coincidence that as we accelerate our peace meditations and prayers, we also see "miraculous" political transformations taking place in Eastern Europe and elsewhere?

Could the popular fable of the "hundredth monkey" be coming true? The theoretical and experimental work of Rupert Sheldrake suggest that both animate and inanimate objects communicate with their own kind across time and space to reinforce old habits, or to spread new ones when appropriate. Sheldrake, a biologist and a philosopher of science, has also proposed experiments to test an even more radical concept: that the familiar "laws" of nature may consist of deeply ingrained habits rather than of "Platonic" absolutes, and may therefore be subject to revision. This concept may well become one of the most profound working hypotheses of the New Science.

The notion of mind-over-matter on a large scale seems appealing at first, but it takes little imagination to see that the technologies associated with the New Science could literally move mountains and must be used wisely.

The New Science, therefore, requires the highest sense of values for its practice. No longer can we afford to create technologies whose side-effects often outweigh their benefits, as we have seen with nuclear energy, allopathic medicines, non-recyclable materials and so forth. The wholeness approach inherent in the New Science also demands an alignment toward some global ideal—perhaps an analogue to the Apollo program, except this time it would be aimed at the survival and enlightenment of humanity itself.

The late Joseph Campbell perhaps best expressed the need for a New Science: "The mystical theme of the space age is this: the world, as we know it, is coming to an end. The world as the center of the universe, the world divided from the heavens, the world bound by horizons in which love is reserved for members of the in-group; that is the world that is passing away. Apocalypse does not point to a fiery Armageddon but to the fact that our ignorance and our complacence are coming to an end."

In the following chapters we will look at theories and successful experiments that may indeed jar us out of that ignorance and complacence. Meanwhile, what better way to end this chapter than with another excursion into the dictionary. Seeking the definition of the word apocalypse, we find its true meaning to be not "cataclysm" but "unveiling!"

3

The New Engineering:
Mind Over Matter

October 1990, Princeton, New Jersey

How strange I felt walking from the parking lot toward the engineering quadrangle. Memories flooded my mind, giving rise to a bittersweet feeling of melancholic nostalgia.

The spreading maples and elms of Princeton University emanated a familiar scent that evoked memories of my time there in the Physics Department more than a decade earlier. It was mid-October yet the air was still sultry, the trees and grass deep green. Disappointing for someone who had just arrived from the furnace of Phoenix, expecting to experience a crisp, vibrant autumn.

As I strolled up to the first building, intending to ask a student parking his bike for directions to room C-131, the PEAR lab, I flashed back to my days on campus between 1975 and 1981 as researcher in outer space development. Along with my colleague the late Gerard O'Neill, author of *The High Frontier*, I had proposed using the materials from the asteroids to build huge solar power stations and colonies in space. This and other futuristic concepts had been only briefly within the acceptable range of funding by NASA, but for years thereafter they continued to stimulate innovative thought and discussion at the biennial Princeton Conference on Space Manufacturing.

My reverie was too engrossing, for the student had parked his bike and disappeared. As I waited for another, I gazed across the campus at the Woodrow Wilson School where we had held those conferences that attracted top researchers from around the world. In 1979, Princeton's Dean of Engineering Robert Jahn, had made an appearance at the conference, at which he had been an organizer and speaker. Jahn was an internationally acknowledged expert on electric propulsion systems and author of the authoritative text *Physics of Electric Propulsion*. While not close colleagues in this conspiracy to colonize space, we had shared a common interest: mapping out how humans could go to Mars and other interplanetary destinations. Although it was a bit far-out for some folks in the space sciences, the work of O'Neill and myself had been blessed by Princeton and by NASA.

In 1979 Jahn and I began to develop a new mutual interest, although we did not learn of our commonality until several years later. The subject was psychokinesis, and it was so far outside our left-brained aerospace view of reality that it would take several years before either of us felt comfortable speaking about it in public. We were "closet parapsychologists," afraid to reveal ourselves to the skeptical frowns of our Princeton colleagues. Nevertheless we began, independently, to explore inner space; it was so intriguing and had such a siren's call to our thirst for understanding that we simply had to heed it, even knowing that the world would look on in disbelief if it were disclosed.

The fountain and pillars of the modern Woodrow Wilson School, set amid Gothic architecture and ivy in one direction and a functional engineering quadrangle in the other, were reminders of those innocent times. I stretched my vision across Washington Street to the ivy-covered building where I had taught Physics for Poets. The comforting sights and sounds of the Princeton campus in 1990 were not much different from those of a summer eleven years back, but my feelings and my goals were very different. The trees, buildings, roads, pathways, cars, professors and students seemed to be frozen in time. Or was it I who was warped in time, a stranger in a strange but safe land, the womb of Academe?. . . Or was it safe?

As I asked another student for directions, my pulse elevated lightly in anticipation of rejoining Jahn for the first time in more than a decade, during which time each of us had taken on greatly altered identities. Each of us, in his own way, had come out of the closet. Since I was no longer affiliated with Princeton or any other major institution, a part of me felt as if I didn't belong here, even as a visitor; that I was an outside-the-box carpetbagger, no longer a professor, no longer an insider. Jahn, at least, had kept his institutional credentials intact.

How different things might have been if I had known what he was up to in 1979 and had joined forces with him, I thought, or if I had simply stayed on in conventional science.

The PEAR Lab

The basement corridor of the engineering school suddenly dead-ended in a sweltering machine shop. Was this the PEAR lab, the psychokinetic hub of the universe? No, it wasn't. A machinist courteously directed me back to the corridor, then to an insignificant door which could have been a closet. This was indeed the PEAR Lab, Room C-131.

PEAR stands for Princeton Engineering Anomalies Research, a curious title which tips us off that what goes on within may not be "authorized" under the "reality rules and regulations" of current science. The impression becomes clear that we may be going outside the box as soon as we enter this laboratory ignominiously stashed away in the basement.

Anomalies. Experiments in harnessing electricity and

magnetism are not regarded as anomalies. Neither are computers, nuclear power plants, rockets, jets aircraft, or even a relativistic linear accelerator or quark bubble chamber. But just-as-real psychic energies are thought of as anomalies. Why?

The word "anomaly" (a "deviation from the common rule") is often used as a diplomatic label for the possibly paranormal that does not necessarily certify its acceptance. There in that basement corridor I paused to ponder the scientific illegitimacy of mind over matter that still necessitated such diplomacy.

I opened the door to a small room crammed with an overstuffed couch on the left, a driftwood coffee table in the middle with sculptures of pears (the fruit) on top, and Murphy on the right. Murphy resembles a vertical pinball machine ten feet tall and six feet wide with 19 bins across the bottom containing 9,000 polystyrene balls. Murphy was to be my first contact with the paranormal at Princeton.

The other four rooms were also crammed with desks, computers, electronics, books and bookcases, notebooks, printouts, and Princeton tiger paintings on the walls: the creative accoutrements of an eccentric intellectual, perhaps advised by an interior decorator from Tijuana.

My son Brian greeted me. Brian is an undergraduate at nearby Rutgers University and volunteers for some of Jahn's experiments. He was there at the moment, of course, by prearrangement.

I seated myself in the overstuffed couch. No sooner had I sunk my fifty-year-old body in deeply than a veritable parade began to pass through the lab: researchers, students, administrators, visitors, sponsors, and, finally, Jahn himself. Tall, lean in body and face with concave cheeks and thinning hair, and wearing a jacket and narrow black and orange Princeton tie, Jahn, 65, paused for a moment to reintroduce himself.

"Good to see you again, Brian.... I'm sorry I can't

see you now," he said. "As you can tell, it's a busy day here
and I'm now between a faculty meeting and introducing our
Soviet visitors to the President. We also had unexpected visitors from Fetzer." The Fetzer Foundation of Kalamazoo,
Michigan, was created by the recently deceased former own-
er of the Detroit Tigers, John Fetzer. It was one of the foun-
dations sponsoring Jahn's research.

"We'll get together later," Jahn continued. "Will you
still be here at seven?"

"Yes," I said, feeling some awe for his popularity and
some sympathy for a man whose life was inundated by curious supporters and vociferous detractors. My own busy life
seemed tranquil in comparison.

Shortly after, Brenda Dunne introduced herself. She
is Jahn's research collaborator and co-author of their 1987
book , *Margins of Reality*, which describes their work. Warm,
personable, casual, articulate, with a Latin-looking round
face and hair swept back into a braid, Brenda started to tell
me about herself.

"I'm a developmental psychologist from Chicago,"
she began. "Some say this is a good balance for what Bob
began here. One wouldn't have thought me to be in an engineering department but somehow it all works out. The engineers design the equipment and run out the numbers while
I take care of the human element."

"Probably unprecedented for an engineering department," I said.

"As far as I know, this is it." She continued, "You
know, one of the reasons many earlier efforts in parapsychology may not have succeeded is because they didn't have
interdisciplinary teams, able to compile a large quantity of
experiments that can be repeated over and over again. The
sheer volume we've compiled over ten years is the kind of
thing that's needed to establish a credible database, one
we're eager to see replicated outside as well as inside the
PEAR Lab."

Just then, Dunne was called back to action in the lab. Rarely have I seen such activity and excitement in a university, and in such an unlikely place as a former basement storage area. Within minutes I had held conversations with such diverse individuals as Jahn, Dunne, the Director of the Soviet Institute for Theoretical Studies, students, researchers and sponsors. I had found myself exchanging calling cards, autographing all four copies of *Exploring Inner and Outer Space* I had brought with me, and wishing I had brought more. I was in the right place all right, and the melancholy of my Princeton memories disappeared, leaving only excitement and satisfaction.

Running Murphy

With both Jahn and Dunne gone it was now time for me to perform some experiments. Still sitting deeply in the overstuffed couch, I gazed across to Murphy eight feet away. "May I do an informal experiment with Murphy?" I asked Angela Thompson, one of the researchers at the lab and my temporary hostess.

"Sure, but you need to realize just one trial doesn't count, that we would need to run some controls too for this to be in the database. "

"That's O. K.," I answered, "this one'll be off the record." And so, with a crash, the 9,000 balls emptied from the nineteen bottom bins and gradually ascended along a conveyor belt to the top center. From there they would be dropped squarely onto a peg with a fifty-fifty chance of going either left or right, then onto another peg at fifty-fifty, and so forth until they fell into one of the original nineteen bins at the bottom. It would take all 9,000 balls fifteen minutes to fill the bins. This mechanical cascade filled the lab with the sound of a pelting hailstorm.

Meanwhile, Brian Jr. and I sitting on the couch "willed" the balls to go to the right.

This is a good experiment for the lay skeptic, I thought. Here it was, a rudimentary Newtonian mechanical device, bared in all its glory, concealed behind no mysterious electronic happenings inside of a black box, as in the case of Jahn's random event generator. How ironic that a purely Newtonian device might be what was needed to demonstrate the non-Newtonian reality of mind-over-matter to our culture. And how utterly simple the device was, the antithesis of a billion-dollar particle accelerator or space telescope or fusion reactor of the sort that would occupy most Ivy League engineering or physics minds. Now I could begin to understand why some of our colleagues may have felt Jahn had gone mad. But if madness is next to genius, others, myself included, saw only genius in action as the pinball machine noise ricocheted through all five rooms of the PEAR Lab and the bins began to fill up.

If Murphy were suitably zeroed or calibrated, and if my son and I were not exerting our influence on the balls to go right or left, we could expect them to fall into a bell-shaped pattern, with the highest number of balls falling into the center bin. This is how the balls appeared to us before our experiment began. As Brian Jr. and I saw the balls trickle down, we gave the right side our body English and our mental and verbal focus. "Go right, go right!" we chanted.

Very soon we started to notice a change in the distribution: more balls now did go toward the right. When the 9,000 balls and all the noise finally settled, a polaroid camera on the wall above us captured a picture of the balls (See Figures 4 and 5).

"Very odd," Dunne said later. "I don't think I've ever seen one of these, where the balls in the center bin are actually lower than those in the adjacent bins and where the balls went that much to the right. And look at the bin that's two over to the right from the center, almost as full as the center bin itself and way above the corresponding bin to the left. But of course we don't have a formal calibration so we can't include this run in our database."

This was disappointing, but fair enough. It was also an early indication to me, later confirmed by a look at their data, of the extreme caution with which the PEAR Lab approaches its experiments. All research is tightly controlled. Credibility is the key word, an answer, on the skeptic's own terms, to the skeptical challenge, "extraordinary claims require extraordinary proof." But Jahn's PEAR Lab uniquely joins the two worlds of Western science and metaphysics, seemingly so far apart and yet now so close. That's why so much action, excitement and stimulation could fill the air in what would otherwise be a rather stuffy, academic setting.

The Random Event Generator

As I got up from the couch I once more looked at the lopsided Murphy, feeling a victory of sorts, albeit one that I could easily agree didn't count scientifically. Now, awaiting me in the room to my left was the Random Event Generator, or REG (Figure 6). This is the sophisticated version of the original student project which—because the student was willing to follow her urge to know the truth—had so changed Jahn's life. Designed to zero-out all but random variations, this tabletop black box about the size of a personal computer is able to spit out 200 random binary numbers each second (high or low, plus or minus, heads or tails, ones or zeros). Price tag for building it: $50,000. The REG can do electronically the equivalent of flipping an unbiased coin more times in one day than has been achieved in a lifetime of parapsychological experiments. The PEAR Lab scientists could satisfactorily pick up effects as small as one part in ten thousand after a few hours of trials.

The experimental protocol was very easy to learn. I simply sat down with the machine running a series of trials, and would think high, think low, or think neutral about the machine. Each trial would last for two minutes. I could do whatever I wanted to prepare for a trial, then I would declare my intention in a given trial by picking one of three conditions (high, low or neutral). Then I would use whatever technique appealed to me to coax the numbers from a neutral 100 upward (signifying a dominance of the higher binary numbers) or downward (showing a preponderance of lower binary numbers) or not at all. I did thirty trials, having selected in my own order ten trials of each of the three conditions. Throughout each trial the results were displayed to me in large red numbers on a monitor, one number every second, showing my cumulative score meandering above or below the random mean value of 100.00 while I was doing whatever I could to bring them up, bring them down or keep them near 100.00. Running the experiment felt odd even to me, as accepting as I already was of the work. I realized I was self-conscious.

Surprisingly, I had some strange results that were difficult for Dunne to interpret. They weren't random but they were not necessarily in the direction I wanted for them. The statistical spread of numbers was not at all normal. This could well be my individual "signature" but more runs would be needed to find out.

Things take time. Such is the nature of science.

"That's why we've now accumulated over a million trials run by 40 operators over the past ten years," she said. "We're primarily interested in getting a large statistical base."

What have the PEAR scientists learned from more than a decade of data collecting? First, that there is unquestionably a mind-over-matter margin of reality in the experiments. The probability of the observed deviations being that far from random are about one in five thousand. The effect is small: most "operators" (test subjects) get the result they want (high or low) by only one part in a few thousand over chance. In other words, when they think high, their cumulative score is about 100.04 and when they think low, their score is about 99.96, with 100.00 for the neutral runs. Small numbers, but the experiments show virtually everybody has psychokinetic ability to some degree.

Second, it seems to make no difference whether the operator is present in the room with the machine in Princeton, or down the street, or in California, New Zealand or Europe.

Third, each operator seems to have his own unique "signature" or graphically represented expression of his non-random interaction with the REG machine. These signatures are repeatable enough to reliably distinguish one operator from another. The implication is that we all generate our own margin of reality.

On the basis of these observations we could imagine building electronic devices tuned to our own mental signature—for example, a garage door opener (which we would want to make sure is also tuned to our spouse's signature!). The design of such devices would involve electronically throwing out the random component and tuning in only to the psychic component. Of course, we would have to learn more about whether we can pick up this effect in a matter of seconds, not hours. We might also have to train ourselves to use these "psychic muscles" in a more powerful and effective way. Also, as we shall see in Chapter 7, we can find innovative ways of analyzing the data to obtain much larger effects.

"How about having thousands rather than hundreds of numbers generated each second?" I asked. "Can you shorten the necessary time to obtain significant data?"

"We tried that," answered Dunne. "It doesn't seem to matter. We still start getting significant results in a matter of minutes to hours for most operators."

"Have you tweaked any of the variables? For example, have you sought out psychically-gifted operators?" Dunne's answer was that they have drawn their operators randomly, from both student and general populations.

"Do you know how the results correlate with the personality of the operator?" She smiled knowingly about all these frequently asked questions. "No," she said. "That's for somebody else to do. Our job is to compile the database, to make it credible so we don't have to deal with anomalies on top of anomalies and hand the skeptics even more ammunition. Most importantly, we throw no data out; some operators have signatures just opposite to what they want, but of course, that too is significant."

As as a result of my discussions with Dunne, I found myself wondering increasingly whether or not we had unconscious mechanisms that routinely sabotage our psychic abilities, perhaps because they're socially and scientifically unacceptable. And if so, could we somehow coax or trick the saboteur out of the way? Can we consistently produce effects greater than the slim psychic margins of reality observed at Princeton? Later in this book we shall see that these things are indeed possible and that the PEAR experiments are just the tip of the psychic iceberg.

As Dunne once again excused herself, I went back into a reverie about my own scientific past. Suddenly it occurred to me that I may have sabotaged some of my own scientific work years ago. Could my psychic abilities have created erroneously positive physical results because I willed it that way? At the time it would have been inconceivable, but now it as worth taking another look.

My Earlier Princeton Days with the Physicists

I was once again brought back to my days in the Princeton Physics Department, and could remember all the talk about how ridiculous claims of the paranormal had been. At the time I had been denying my own anomalous experience and Jahn had quietly begun his experiments.

I recalled talking with Phillip Anderson, the Princeton physicist who received a Nobel Prize while I was there in

1977. I also remembered my conversations with other Princeton physics Nobel Laureates John Wheeler, Eugene Wigner and Val Fitch as well as about fifty others, all men, at the biweekly Joseph Henry luncheons in the Physics Department building. Debunking parapsychology had been commonplace at these events. I participated in these discussions while sipping thimble-sized glasses of sherry under Joseph Henry's portrait and feeling rather smug about my scientific certainty about how the cosmos operates. Henry had been a nineteenth-century pioneer in uncovering the anomalies of electricity and magnetism, and gazing at his portrait I had often wondered how his ideas may have at first been received. In 1977 none of us had known that one of our own kind over in engineering (the Dean no less!) would be directly flouting the orthodox consensus. Another Joseph Henry in the making?

In a recent interview published in *The New York Times Magazine* and reprinted in *Omni*, Phillip Anderson stated that Jahn's work must be in error, because if it were not, the laws of physics must be revised. It was this point, I reflected, which made the skeptics so confident in their own belief that nothing could shake up the sanctity of physics as it now exists. "Most of the senior members of the physics department did not think this [Jahn's experiments] was a good thing to be going on," said Anderson. Shades of my old Joseph Henry luncheon experiences, I thought. "When Bob calls himself a physicist," Anderson said, "we're a bit skeptical, because if he is right, he would be negating the entire basis of the profession." Anderson wondered why his own precision experiments weren't sabotaged by his thoughts or why people weren't trying to break the bank in Las Vegas.

In my opinion, Anderson was both right and wrong. He was right to recognize that physics will need to be revised. He was wrong to assume that Jahn was incorrect, and that this particular kind of mind-over-matter effect could necessarily influence the small statistical samples of wagering (and the particular mechanisms employed) in Las Vegas. He was also wrong in assuming that new knowledge necessarily "negates" the old; after all, Einstein did not negate Newton, but simply placed him in a larger context. Anderson could be right or wrong about his own experiments possibly being biased because of the mind-over-matter effect. For one thing, the effect may well have been too small to have made a noticeable difference. Or the experiments could in fact have been biased without his yet knowing about it!

Jahn himself had once become interested in the possible effects of the mind on sensitive electronics in aerospace systems. This was in fact one of his early motivations to do the PEAR Lab studies in the first place: if you're looking for physical effects, you could sabotage experiments with your thoughts. If what you are looking for is a mind-over-matter anomaly, you could sabotage the results through a scientifically orthodox mindset. Either way you could lose if you were not aware.

Anomalies in My Own Research

All this prompted me to look for anomalies in my own earlier work. Did my thoughts influence the results of any of the many experiments I conducted during my scientific career? One set of observations was immediately suspect. I recalled that as a graduate student at the University of California at Berkeley in 1966, I had begun to look for signs of ice crystals in the atmosphere of Venus by using light-measuring equipment on a telescope. This was my first original research project that contained the potential for an important discovery. Ice in the Venusian clouds could mean Earthlike conditions and perhaps even the presence of life lower in the atmosphere. I had been highly motivated to find the ice.

The premise of the experiment was simple. If hexagonal ice crystals were in the cloud tops, we could expect

Venus to brighten when, in its orbital motion, it passed through a Sun-Venus-Earth angle of 158 degrees. This brightening would correspond to the familiar halo we often see as a ring 22 degrees from the Sun or Moon. We know that the halo arises from sunlight double-refracting and reflecting from hexagonal ice crystals acting as prisms. The 22-degree halo in our own atmosphere is a telltale sign of the presence of ice crystals in a thin cirrus cloud layer. Could the same be true of Venus? Placing a sensitive light-measuring device called a photometer at the focus of a small telescope might give me the answer. Some of my professors were also enthused about this project and gave it their blessing.

And so off to Kitt Peak Observatory in Arizona I went, prepared to measure the brightness of Venus as a function of time to see if the halo effect would appear. I was very excited about it, and felt all the intense involvement of an idealistic young scientist pursuing his first creative exploration. It might even yield a quick Ph.D. thesis!

But I restrained my enthusiasm and conducted myself as objectively as I could. I wanted the photometer to reveal the truth, whatever it would be. During a day's observation I was pulling for levels of brightness that corresponded to my theoretical model for the presence of ice. At the same time the experiment had numerous controls to eliminate the chance of instrumental bias: on a given day I would move the telescope back and forth again and again between Venus and reference stars to take out systematic errors. The random scatter of the sky brightness fluctuations would be the largest source of error.

(Scientists often talk of two kinds of error in their measurements: random error and systematic error. Random errors are relatively easy to deal with statistically and systematic errors are usually a result of faulty design of the equipment, sloppy experimental procedure or lack of controls.)

In 1966 I did find a halo effect, although it was frustratingly marginal. The magnitude of the effect was about the same as the random observational errors, and so the best I could do was to set an upper limit to the amount of hexagonal ice crystals in the Venus cloudtops of a few percent. I published the results in the prestigious *Astrophysical Journal* but was told not to do any more work on this for my Ph.D. thesis, in part because the results were too inconclusive. The project was put on hold.

But three years later I had another opportunity and took full advantage of it. Having completed my doctoral work and having finished a seven-month stint in the scientist-astronaut program, I returned to planetary astronomy as a faculty member at Cornell University in 1968. The next year I flew to Cerro Tololo observatory in La Serena, Chile to take advantage of the next pair of Venus halo opportunities (which happen every nineteen months). This time the weather was perfect, the equipment was well-tested, the controls were tight, and I was more experienced, confident and mature. I felt truly encouraged to give it all my best.

This time I found the halo! As it passed through the appropriate reflective angles on both sides of the sun, Venus brightened where it should have, and fit my theoretical model very well. I found an amount of brightening more than twice the extent of random errors on each day of observation, giving a high degree of credibility to the presence of at least some hexagonal ice. The probability of this being true was greater than about 98 percent. I submitted my research paper to another well-known journal, *Icarus*, and it was praised by my peers who reviewed it.

Still, I was cautious about my interpretation that ice condensed in the Venus clouds. Too often in science have investigators zealously "found" what they were looking for only to find later that it wasn't there. (This, of course, is Anderson's belief about Jahn's work.) So I concluded that the observations suggested that at least a portion of the cloudtops might be composed of hexagonal ice but the effect was still marginal and needed confirmation.

That confirmation never came. In the years that followed, other observations revealed sulfuric acid to be the primary constituent of the Venus cloudtops, and that the temperature and pressure conditions would not be right to form hexagonal ice crystals. Another attempt I made, with a Cornell graduate student, to find the Venus halo showed no halo at all. We published the result in *Icarus*.

Whatever happened to the Venus halo? Did the ice crystals only show up in 1969 and maybe in 1966 but not during the 1970s? Did some unknown systematic error crop up in the earlier observations? Or did my mind "will" the results I wanted, as I enthusiastically chased my own form of rainbows in exotic Chile? In other words, did the photometer respond to my wishes while I felt I was on a clear path of discovery, and later dampen because a new consensus reality emerged that excluded the possibility of ice in the Venus clouds?

The PEAR Lab experiments suggest to me that latter alternative may have been what happened. The transient presence of ice in the atmosphere under physically impossible conditions, or a series of unknown systematic errors that could so well mimic a Venus halo effect, seemed even less likely. At that time I had rejected my own earlier observations and, not knowing about the mind-over-machine effect, I was dumbfounded. I had thought something was wrong with me.

Now, more than two decades later, I suspect that my deep mindset created the halo effect of 1969. It may be that the electronics had responded not only to the light coming from Venus but also to the energy coming from my mind. And if that is so, how many scientists have likewise inadvertently created this same effect-error how many times? Dozens? Hundreds? Thousands? Millions? In the future, scientists may need to "think neutral" going into sensitive measurements of physical phenomena.

What Jahn is trying to tell us is that these kinds of circumstances are the very ones that address Anderson's accusation—that parapsychology experiments contain an internal bias that "makes people in these fields (parapsychology) cheat either consciously or unconsciously." Jahn is indeed aware of the effects of bias; however, he has isolated them and acknowledged their importance as a new and radical dimension of the scientific research endeavor itself, rather than reducing them to a form of cheating.

Isn't it ironic, I thought, that the very thing that may have sabotaged my efforts to obtain a clear answer to the composition of the clouds of Venus is the thing that interests the new scientist in me most. Old scientists see it as unconscious or conscious experimental bias at best and cheating at worst—to be avoided at all cost—while new scientists see it as an opportunity to learn about some of the greatest mysteries of our time: How does human consciousness interact with the environment? And to what extent do we create our own reality?

Robert Jahn and the New Science

Jahn returned to the lab at about seven in the evening, weary but still a man of his word about our getting together. It was pretty clear he hadn't had a break for hours. . . and there was yet more to go for him. The main item on my agenda for this visit was to see if he would be interested in joining the core group of our newly forming scientific organization.

"I have to be careful of my commitments," Jahn told me when I asked him to consider playing a role in the International Association for New Science and its related research institutes. "I'm already active with the Society for Scientific Exploration which I believe provides the appropriate forum for reporting our work."

"The SSE looks at these things as anomalies," I said.

"In our group we are looking at a new paradigm for science so they are no longer regarded as anomalies, as exceptions to the rule."

He looked at me for a moment. "Yes, we do need a new paradigm...."

Jahn went on to explain that he was already over-committed and had his name on enough mastheads. While he could make no guarantees, he did ask to be kept informed about what we at the IANS were doing.

As Jahn left for his next appointment, dutifully fulfilling commitments into the night, I reflected on how difficult it would be for a man with so much responsibility and respectability to pause for long enough to move yet further out of a box which had for so long nurtured him. Here was an extraordinary man often caught in a crossfire trying his best to balance all these factors. I had to admire him.

That night I was exhausted from overstimulation. After saying goodnight to Brenda Dunne at eight-thirty, my son and I returned to the motel where I fell asleep immediately. My last thoughts of the day were that the carefully controlled PEAR lab experiments now provide all the engineering validation we need to establish beyond any reasonable doubt the existence of a mind-over-matter aspect to reality.

These very important results form a foundation for further legitimate experiments in what we have called the paranormal. They do not necessarily validate all such efforts, but they do require that we now be willing to think the formerly unthinkable.

We must now rewrite the laws of physics so that the results of such experiments are no longer anomalous. In doing so, we must never forget that the "laws" we write are simply reflections of our own current understanding of reality. They must never be confused with reality itself, which is always greater than the words and concepts we use in our attempts to manage it.

Robert Jahn's remarkable experiments establish a priceless link between an old and a new science. The effects he and his colleagues have observed, though small, are very real. They are keys for unlocking a narrow passageway to the wonders that await us outside the box.

Some of us might be more impatient about making that passage. Blatant demonstrations of psychokinesis can quickly widen that passageway without the need for sophisticated analysis to screen for small effects. For those so inclined, I recommend a journey to Puoso Allegre in Brazil, where one man repeatedly takes us outside the box with his clear demonstrations of mind-over-matter—this time at the one-hundred-percent level.

4

Ra!

Stretching our legs in the parking lot of the Fernandao Hotel was a relief from the bumpy six-hour ride in a cramped car. I had a day off from the course I had been teaching on New Science in Belo Horizonte, a clean, modern city of three million about 300 miles north of Rio de Janiero in the rich mining and agricultural state of Minas Gerais.

I reflected with some apprehension on our imminent appointment with Thomaz Green Morton, perhaps the most gifted "breaker of the scientific rules" in the Western world. We had given up any hope of seeing him until the previous afternoon, when he had changed his mind and decided to see us rather than to travel to Sao Paulo for a television appearance.

The appointment hadn't come easily for my Brazilian host, Walber Pinto of the Center for Esoteric Studies. He had called Thomaz on my behalf but was rebuffed on several occasions, being asked for his credentials, reasons for the visit, whether a stiff payment might be appropriate, etc. Thomaz had then handed the matter over to one of his disciples, Maria Celia Teixeira, a Brazilian homeopath with whom I had been unsuccessfully negotiating over the telephone during the previous few days for the privilege of seeing Thomaz. Then the way to the master of the paranormal suddenly and mysteriously cleared when his secretary called us with an invitation just the day before.

What would Thomaz be like? What miracles would he perform this evening? Or would he decline? Would we see him for just a brief interview and then head home? Or would we have an hour? Or two? Would we come across simply as gawkers waiting to see Thomaz perform his incredible feats, with no interest in his human side? Could we capture his feats on camera? We would soon find out.

Some synchronicity had led to this event. Just two months earlier, Walber had unexpectedly invited me to come to Brazil to teach a one-week course. Within days of that invitation I received in the mail a copy of a just-published book, *Miracles and Other Realities* from one of its authors, Lee Pulos, a psychologist from Vancouver, BC. The book was a biography of Thomaz, and it totally captivated me. It described how Thomaz astounded visitors by performing some of the most amazing feats imaginable, ones that seem to make a mockery of the materialistic world view. Pulos gives eyewitness accounts of Thomaz' ability to materialize and dematerialize objects, mend them, bend them, break them, transform them and transmute them, as well as his unpredictable inclination to perform healings, disappear, reappear, bilocate and produce flashes of light from his solar plexus area.

For those fortunate enough to experience Thomaz over the years, being in his presence provided an immediate and inescapable paradigm shift. I wanted to be one of them.

Why would anyone need to do this if he were already convinced there was something to Bob Jahn's experiments? Because for some reason our rational, deterministic left brains don't seem to have the power to convince us to accept miracles in our daily lives. Reading a good book or performing a successful experiment on psychokinesis may help. But that only seems to whet our appetites. Having isolated anomalous experiences such as a remote viewing, a near-death experience or a healing may also help, but these can seem to fade in time and are difficult to "ground" in science. But to actually see a gifted psychic perform transcendent feats repeatedly and at close range over the course of an evening is certainly convincing to all those present, and the memory never seems to fade.

To fully come to grips with our potential for mastery over the material world is to undo or revise much of our formal scientific learning. As a trained scientist I have needed to do a lot more unlearning than most people; nevertheless, I realize that I am farther along than most of my colleagues—particularly the mechanists who have organized themselves into debunking groups that steadfastly deny all paranormal phenomena and toe the line of scientific orthodoxy, much as the Flat Earth Society in England is reported to have done in its heyday.

In the end, there was no substitute for being there. As seductively convincing as Pulos' biography of Thomaz was, the cautious skeptic within me had to see it all firsthand. These were my thoughts while we awaited Thomaz.

A few minutes before we were to be taken to Thomaz' house, a car suddenly swerved into the hotel parking lot. It was Marie, with a journalist and photographer. Marie enchanted me. Short, fortyish, with long black hair and bangs, she was brimming over with enthusiasm as she yelled the Thomaz trademark "Ra!" (pronounced in Portuguese as "Ha") and hugged each of us as if we were intimate friends. For us tired riders, the energy shift was overwhelming and unsettling. This was a foreshadowing of the charismatic, magnetic evening to come—one whose emotional impact was to affect us as much as the objective reality of the paranormal performance we were going to observe.

"See!" Marie exclaimed in good English, "I said I'd be here in time! You're here early! We still have a half-hour before your appointment and we're only five minutes away!" She grabbed my hand. "Let's go to where it all happened, where it all began! Do you want to come? Say yes! Let's go!"

Without a word we hopped into the cars. Marie was taking us to the boyhood home of Thomaz where, on his twelfth birthday while fishing nearby, he was apparently struck by lightning or, as the story goes, some "cosmic ray" that seemed to give him his psychic powers. So we all drove out to the countryside and walked the land of "where it all happened"—the farmhouse and church of his youth and the small pond where he was fishing at that fateful time. Nearby was a recently constructed shrine of the Virgin Mary where many of Thomaz' miracles were said to have taken place, shades of Fatima and Medjugorje. We posed for pictures, and the journalist interviewed me as a visiting celebrity in that remote Brazilian town.

It was now 5:15. Tropical dusk rapidly descended on the landscape and we departed for the more opulent house of Thomaz a few minutes away. Driving through a gate into a courtyard, we waited for our audience in a small guest house with a large sign "Ra!" over the entrance. Dusk soon turned to darkness, and it was a chilly southern Brazilian winter evening as we were escorted to the main residence to meet Thomaz.

Another overwhelming encounter. Thomaz exuded a macho charm and charisma I hadn't seen since the days I had spent with astronomer Carl Sagan over twenty years ago. He even looked a bit like Carl but that's where the resemblance stopped: their belief systems about reality were surely universes apart.

Thomaz and I hugged tightly, and we ceremoniously posed for pictures and exchanged books and other memorabilia. The room exuded a perfume scent, one that was said to have been materialized by Thomaz. Then we walked out to the cool porch and listened to his nightly six o'clock meditation. For about thirty minutes we heard Thomaz exclaim powerfully his thanks to God for the stars and sun and moon and earth and trees and streams and lakes. Walber translated all this into English, as about ten of us (some were in

Thomaz' entourage) sat on the porch facing the stars with heads turned toward the left where Thomaz was sitting on a hammock lit up by lights enough to be videotaped while the rest of us sat in darkness. I later learned that a man often followed him around with a video camera to record his remarkable life.

The meditation was a daily routine at six o'clock Brazilian time. We were told we could tap into this energy any day at this time and miraculous things could happen. A sincere, calm and intelligent older lady talked to me in good English about Thomaz' recent spiritual evolution and the work they were doing for personal and planetary healing. I was sympathetic but also uneasy about the incongruous glitter in this rural setting.

Miracles in the Restaurant

When the meditation ended I realized we had two possibilities before us: we would drive home unfulfilled or we would be invited to stay for an evening with the amazing Thomaz. The news was good: he invited us to join him for dinner at a local restaurant, and it was there the action began. We took the short ride down the hill to one of the only two restaurants in Puoso Allegre that Thomaz frequented—unsophisticated, and brightly lit by fluorescent lights. About a dozen of us sat around a long table in a large room with a high ceiling. The casual atmosphere resembled that of a rural truck stop in America, a reflection of Thomaz' humble origins.

Thomaz ordered a bottle of rum to mix with his coke, and we began to eat. He sat directly across the table from Walber and myself. All attention and talk were riveted on Thomaz. His charisma and the atmosphere of expectancy were overpowering.

Thomaz drank a couple of rum-and-cokes, which seemed to loosen him up a bit. He then felt a surge of energy, closed his eyes, and yelled "Ra!" He then extended his hand about a foot above mine and perfume began to roll out of the back of his hand and down his fingers and drop into my hand (Figures 7 and 8). I looked more closely at the back of his hand, my eyes were about six inches away from the aromatic fluids rolling out of his hand. What I saw I found hard to believe, and though I masked my denial with laughter it was combined with a childlike sense of awe and wonder. At that moment I was hardly the image of the objective Western scientist-observer. Perfume was coming out of his skin. This was no sleight of hand.

Soon perfume was oozing from his hand in sufficient quantities to pool onto our hands and the table. The scent was strong and pervaded the room. During the seconds it took to materialize the perfume, Walber and others remarked about flashes of light coming from Thomaz' solar plexus and into his hands. These flashes evidently marked the surges of energy that made this manifestation possible.

The excitement ran high at this moment. Like children seeing Santa Claus for the first time in a toy store, we all beamed and gasped, directing all our attention toward Thomaz. For some of those present, emotions began to stir up.

Cocky and seeming in control, Thomaz poured himself another rum and coke, stirring it with a knife. Waiters brought on endless skewers of meat which they dropped onto our plates. Having completed about a third of the rum, Thomaz began to speak more loudly now, repeatedly yelling "Ra!" and attracting more attention from others in the restaurant who knew him and came to the table for some booze talk.

He gazed at each of us intensely and invited us to go with him to "Aphron-V," a planet in a parallel universe he claims to have visited repeatedly during those times he has dematerialized himself. This is also where he claims to have received the curious marks on his forehead shown in the pictures in this book. Somewhat surprised by his invitation yet wanting to placate him, Walber and I said "yes," while another said "no" with genuine concern about our apparent gullibility.

It was at this point that some of us began to feel the negative side of the magical solar-plexus energy emanating from a drinking Thomaz. Was it Thomaz himself who was doing this? Was it some kind of energy possessing him that became increasingly apparent as he downed more rum? Or was it all merely a reflection of our own gullibilities and fears? As Pulos often stated, Thomaz himself never meant any harm to anyone. I believed that, but perhaps we were all dealing with something larger than ourselves. Even so, was that any reason to judge or run away?

For us visitors, what at first appeared to be a scientific session involving detached observations of paranormal events was transformed into a challenging series of emotional and spiritual intitiations. For some of us the whole business was outside our comfort-zones; one didn't need to be a mechanistic cynic to feel uncomfortable. I began to protect myself spiritually using a technique I had learned to apply when the going seemed rough.

It became obvious to me at that moment that the practitioner of the New Science has to encompass simultaneously two seemingly contradictory realities. One is that of the detached experimenter, verifying by direct observation the primacy of mind over matter (and, in this case, the mockery being made of materialism!). The other was that of the observer-participant giving up some of his own impersonal objectivity to take part in—become a part of—the event itself (in this case the charismatic vortex, the shamanistic ritual of mood altering subconscious-stirring energies radiating from Thomaz).

The fundamentalists would clearly call this the work of the devil. Thomaz and his disciples would call it "positive" energy. Did the truth lay somewhere between—or perhaps above?

As I watched Thomaz, it became increasingly clear that in such circumstances—ultimately in any circumstances—there was no such thing as a purely objective observer. The price we must pay to enter the theatre of transcendence is to jump in and participate. Yet the materialization of the perfume was still an objective event with tangible results, seen and smelled by all.

It is perhaps the greatest challenge of the New Science to mediate between the seemingly opposed views in this apparent paradox. It helps that I still carry a sample of the perfume with me.

The evening with Thomaz continued with Walber holding a coin in his hands and asking Thomaz to engrave Walber's name on it by the action of a simple "Ra!" (This is a commonly reported Thomaz miracle.) However, going into concentration and again yelling "Ra!", Thomaz did not achieve his goal. Meanwhile one of us had been focusing his energy on nothing happening to the coin—perhaps mental energy could undo mental energy. It was clear that Thomaz was unhappy about this; but just as soon as we all felt the disappointment, another "Ra!" was uttered and a spoon in front of Walber directly across the table from Thomaz suddenly bent at right angles where the bowl met the handle, untouched by anyone, a seeming consolation for the non-miracle of the coin.

All this inspired me to pick up a spoon and try my own luck at spoon-bending. As I'll explain later, I know how to bend and twist a stainless-steel spoon with two hands, using a combination of mind-over-matter and mild physical force to produce results unobtainable by the force alone. (Most of us can learn to do it that way to satisfactorily "prove" personal abilities at psychokinesis; I teach this in workshops.) While I'm no match for either Thomaz or Uri Geller, I did pick up a spoon in the restaurant and quite easily put two twisted loops into it, something that certainly couldn't be done by pure force. Thomaz noticed the final stages of the bending and grabbed the spoon from my hand. In rage, he exclaimed in broken English, "That's no good! You did it by force! That's not the real thing!" He threw the spoon out the door onto the patio.

I felt embarrassed to have tried such a thing. Thomaz was upstaged and he did not like it. I had been warned about this trait of his, but had foolishly ignored it. I had wanted to play, wanted to know, wanted to test. Clearly this was Thomaz' evening.

The evening began to wear, and I was feeling drained and tired. I looked at my watch: it was already eleven o'clock. We had been at the restaurant for more than three hours and we still had a six-hour ride back to Belo Horizonte. We decided to stay on at the Hotel Fernandao and leave the next morning to get back in time for me to speak, something I was not looking forward to in my present state.

Now Thomaz was drinking more and shifting his moods every few minutes. The rest of us were indulgently awaiting his next trick yet also wondering if it were possible with all that alcohol. Then he stabbed a piece of meat with the end of his fork, holding it out with one hand at the end of the handle (Figure 9). He shut his eyes, the energy once again came up and he yelled "Ra!" The fork snapped in two, the end with the meat falling onto the table (Figures 10 and 11). Another miracle!

Thomaz then challenged us that he could read our minds. One of us thought of the word "beautiful" and, sure enough, he picked it up and said aloud "beautiful".

Thomaz' wife Lygia and his son quietly sat beside him, apparently enjoying just another normal evening! I then began to reflect on how challenging his life must be. Here we were, the visitor-gawkers, wanting to see this man for his unusual abilities rather than for who he really was. Many people come to him for healing and put him on a pedestal. How could any human live up to the expectations placed on him, especially given his upbringing? How could he endure these endless dog-and-pony shows? How could he have a sane and normal family life?

Thomaz' life was emotionally confused and he covered up much of that with his macho charisma and a lot of drinking. Not being "normal" on this planet seemed to have its price. Using his special powers to reward his ego, while at the same time having no malicious intent, he received a short-term benefit of a lot of attention directed toward him. But all this also seemed to be killing him. "Power corrupts..." I reflected. But perhaps at a deeper level he was teaching us something. He mirrored those parts of us that have dark feelings and impulses that lie buried in our subconscious. He was stirring our pots for us to take a look at, and for that I'm grateful. I recalled those times in the limelight when people looked toward me for an image, and sometimes all I really wanted to do was to be human—to love and be loved, to cry and laugh, to make mistakes and to be me. Thomaz seems to need the same and so he gets drunk.

Around two o'clock our party began to break up. Thomaz had polished off most of his rum and didn't have a whole lot left in him to give. He walked outside into the darkness about one hundred feet from the back patio door of the restaurant adjacent to our table. As he began to urinate with his back to us, his whole body flashed brilliantly like the lights of a flash camera. One of the flashes formed a brilliant green filamentary structure. When he came back into the restaurant, his body at the table continued to flash several times. I tried to photograph all this, but I'll never know whether I was successful because that film disappeared, as did the journalist's photographs of me and an erased audio tape I had given to Marie. How they were lost, I don't know. I was told that these things happen often in the presence of Thomaz.

And so with Lygia helping him walk out of the restaurant and with a drunk and tired Thomaz handing out candy balls to each of us as his last act of expansiveness, we walked back to the cars into a dark night that called for a long shower at the hotel and a lot of reflection.

We had traveled a long way to see for ourselves a man who has the power to break some of the laws of an old science that had so influenced our lives and dictated our reality. We could not deny what we saw, and science would need to be revised to accomodate these phenomena.

Neither had we bargained for having to deal with the the formidable energies behind this power that emanated from Thomaz' stomach area (what the Eastern mystics call the "power chakra") rather than from heart, which traditionally would have been characteristic of healing energy. Our emotional and spiritual selves were moved beyond our abilities to be purely dispassionate observers; we could not avoid being participants as well. This had certainly been a challenging initiation into the practice of the New Science.

San Jose, California, November 30, 1990

Walber and I had known another remarkable man, a foremost practitioner of the New Science then in his seventies, whose work clearly came from the heart. His name was Marcel Vogel (Figure 12). As an eager student of Vogel's, Walber (Figure 13) had first met me at Marcel's laboratory the previous November.

Some years earlier, having left IBM as a senior scientist, Vogel founded Psychic Research, Incorporated, to carry on his work free of the biases and pressures of the mainstream. He had already held quite a number of major patents as a crystallographer at IBM, but was now on the trail of new discoveries far beyond the confines of the industrial lab.

We have all heard of instances in which crystals appear to have been used for healing on a more-or-less intuitive basis. But Vogel found that if quartz crystals were cut in certain precise ways and mentally "charged" with a positive "love force" they not only could help heal diseases in a person's body but could in some sense "restructure" the chemistry and other less well-understood properties of fluids. For example, the Sycamore Creek winery hired him to use the properties of his crystals to ferment and "age" wine in mere seconds—a process usually taking months or years. As a bonus, in a controlled wine-tasting, two thirds of the tasters preferred the "treated" wine, which seemed to have enhanced flavor and bouquet.

Is it really possible to bring about such radical changes in the properties of common substances? To find out, I visited his laboratory to do a controlled experiment in which Walber, Marcel and I studied the effects of "restructuring" on the presence of mold in fruit juices.

We used a Vogel device called a "fluid restructuring unit." This is simply a box containing a hollow tube wrapped spirally around a Vogel crystal (Figure 14). Fluids, poured into the funnel at the top, emerge, "restructured," from the other end. The unit is simplicity itself: it has no moving parts and is not connected to any conventional power source.

In this experiment we poured several samples of apple juice and orange juice through the unit. Along with control samples of both juices, which had not passed through the system, we injected mold into each sample in its vial, covered the samples, and then tracked what happened.

After some weeks to months, the results became clear: the samples that had been passed through Vogel's system remained unaffected by the mold, which continued to float on the surface of the juice. In the case of the control samples, however, the mold spread throughout and completely blackened the entire volume of liquid (Figure 15). I was amazed at these results and find it difficult to find an old-paradigm explanation.

From what we now understand of Vogel's work, its implications in terms of food preservaton, water purification, pollution cleanup and so forth, are incalculable. His experiments demonstrate that we have the potential of radically changing the nature of our physical environment through the application of loving energy combined with almost absurdly simple hardware. We have here a potentially transcendent technology born out of the motivation to love.

One day in February, 1991, Vogel phoned me. "Brian," he said in his usual, dramatic tone, "it's very important for others such as you to carry on the work. Remember that love is all there is." A week later he died. Tragically, much of Vogel's work remains undocumented, except for that published in his Psychic Research newsletters and passed down to his assistants and students, and to visitors to his laboratory.

At the same time it is beyond question that we can all tap into these same sources of intelligent power with equally "miraculous" potential results. One experimental "proof" I have already mentioned is spoon-bending, a mind-over-matter technique I teach in workshops with almost universal success. In ways such as this we can all begin to transcend the material world at will and become more comfortable with the "unthinkable."

Bending the Laws of Physics

Fullerton, California, June 1990

I arrived at the Fullerton Church of Religious Science early enough to have dinner with Marlene and Gil Oaks. Marlene was minister of that church and leader of a seminar that would be held that evening on psychokinesis and spoon-bending. Earlier in the year I had been at their church and was impressed with her husband Gil's ability to bend a thick-handled soup spoon with ease.

I wanted to learn how to do it myself and tonight was the time to do it. "It's easy, it's fun!" Marlene declared enthusiastically. "Within three hours you will be doing it." I grinned at her skeptically. Me, bending a spoon? I doubt it, but it's worth a try.

Marlene's approach to teaching spoon-bending (the one I have since adopted for my own workshops on that subject) is based on the Japanese martial art aikido. The practitioner of aikido visualizes bringing in the energy of the universe ("ai") with breath inhalation, mentally transforms that energy to life energy ("ki") at the power point near the navel, and with each exhalation extends that ki to the object to be transformed or task to be accomplished ("do" or way of doing it). After a few breath cycles and with positive intention, we are ready to perform a "miracle," to transform or transcend something in the material world.

To begin, we did a number of aikido exercises to gain confidence in this technique. For example, we tried to bend a partner's extended arm, first when he or she tried to resist through brute strength, and then when he or she did some aikido breathing while extending the ki through the arm laserlike toward a point on the wall. It was far more difficult to bend the arm the second time.

Next we tried walking through someone's out-stretched arm toward a point at the other side of the room, first without focusing on the goal, and then extending our ki to that point. Again, the contrast was striking—the energy exercises of aikido really do work.

We also attempted to lift people, first when they were not focused or "grounded" and then when they visualized their ki extending to the center of the Earth with a heavy anchor. Once again there was no comparison: it was almost impossible to lift an anchored-in person, even though their weight hadn't physically changed. These exercises loosened us up and gave us a chance to build confidence for the spoon-bending.

Marlene made it clear to us that spoon-bending was not an intellectual exercise or a function of the ego. It was in fact quite the opposite: the more we thought about it, the more we tried to analyze it or to show off, the less likely we were to be successful. This was an exercise in letting go, visualizing, being clear in intent, and becoming "at one" with the spoon. In fact, I was later to discover that people who had the most difficulty bending spoons were middle-aged intellectual men—like myself. Here we have a case of a credential role-reversal: little old ladies and children usually have an easier time of it than us overachieving men. Once, during a workshop I gave in Australia, the only ones among the twelve people I taught who couldn't bend a spoon were a congressman from Michigan and an engineer from Massachusetts, both bright, middle-aged men.

When it became time to bend spoons at the workshop, my confidence level was unstable and I was caught between enthusiasm and an overworked mind and ego. I wanted badly to do it, but I also realized I had to be relaxed about it.

And so, holding the spoon in my two hands while staring at the floor near my feet below the spoon, I began to visualize, aikido-style, bringing in the energy of the universe through the top of my head with each inhalation, filling my power point with that energy, and exhaling it into the spoon. I visualized extending my ki into the spoon and becoming at one with it. I visualized the spoon softening and bending as I continued breathing rhythmically like a pendulum.

After about fifteen breath cycles, Marlene then led us into a rapid-fire loud chorus in which we yelled, "Bend! Bend! Bend! Bend!...." This phase of the exercise involved planting the intention to actually bend the spoon. What followed amazed me. About five to thirty seconds after planting the intention, some of our spoons literally softened, a spoon-bending "launch window" opened up for us that lasted for just a few seconds, and many of us immediately began to effortlessly bend and twist our stainless-steel spoons. Just as quickly, the spoons hardened again.

My ego stopped me from taking advantage of the first two or three windows that opened up for me during the first minute or so of yelling "Bend!" My disbelief over the first softenings apparently caused an immediate rehardening.

Then I surrendered. I felt what could only be called love and oneness with the spoon. What followed amazed me and I shall never forget the moment. The spoon stayed soft and I suspended disbelief for long enough to put two or three easy, tight twists into it. Then it hardened, yet continued to maintain some intense heat at the point of twisting, enough to almost burn me. Efforts to unbend the spoon later resulted in the spoon snapping in two.

I began to bend more spoons and then curled the prongs of a fork. These were not tinny pieces that could easily be bent by force, but solid, thick stainless steel. Gil Oaks managed to bend a five-sixteenths inch diameter steel rod into a loop! As the few minutes of the chorus "Bend!" continued, the whole room seemed to be filling with spoon-bending energy. Some of us helped others by standing in front of them and "holding" the energy for them. Most of us weren't able to bend anything simply by looking at it—we used both hands. Yet there was no way we could have done this by means of pure physical force.

This course was Spoon-Bending 101. I discovered I could teach it in less than an hour with more than 90 percent positive results. In other words, most of us, regardless of background, education or degree of physical strength, can discover our own mind-over-matter margin of reality. The bent and twisted spoons are our trophies, tangible physical evidence of the seemingly impossible.

I was thrilled by my new ability. Later that night at the Los Angeles airport I tried to bend and twist a spoon. No problem! In fact, I have rarely had difficulty doing it again anywhere, any time. The only exception seems to be when I am very tired and the handle of the spoon is very thick. Then I can bend it, but may have trouble twisting it into loops. However, for the most part and for most of us, once a spoon-bender, always a spoon-bender.

On one occasion, in front of an audience in Fort Collins, Colorado, I was challenged to bend a plastic spoon that was presented to me by a skeptic in the audience. My inner dialogue about this was disturbing and my own doubts were running high. But I went ahead anyway. Much to my surprise, it worked well!

Even though workshops are the best environment in which to learn to bend spoons (presumably due to the collective energy present), I have been able to teach individuals at home. If you are an open and receptive person, using the techniques I have presented here you may be able to do it on your own. Try it!

What is Going On?

It should be clear by now that whatever pathway you might take toward the realization of mind-over-matter (to follow the formalisms of experimentation, to be dazzled by the performances of a "master" or to be your own master of metal-bending or fire-walking, etc.) you can readily prove to yourself, if you wish to, that something is happening that has no explanation within the Western scientific model of reality. I seriously doubt that anyone but the most hardened, blinded skeptic could deny the authenticity of these experiments and phenomena.

What is going on? The question is not easy to answer, partly because theory is seriously lagging behind observation. What can science do to even begin to answer these questions?

In the seventeenth century, Galileo and Newton began to examine the motions of falling bodies and heavenly bodies. They found that apples and planets responded basically in the same way to the forces of large attracting masses, Earth and the Sun. Newton was able to describe these motions mathematically and called the attractive force gravity. But neither he nor any scientist since his time has been able to explain *why* gravity exists. Although gravity waves and particles have been hypothesized, nobody has actually detected them. All we know about are the effects of gravity. We do not understand the causes. And more recently we have found that even the mathematics are not exact; in fact they become seriously distorted in situations in which the attracting bodies are either extremely dense or are moving at nearly the speed of light.

Similarly, the causative mechanisms of the influence of our consciousness on the material environment are not known. And because the degree and nature of the influences depend on so many variables such as the psychokinetic ability of the experimenter and the properties of the material being transformed, it has been difficult for scientists to quantify, characterize, and control the experiments so a coherent mathematical theory can emerge.

Our problem is somehat like that of trying to understand a black hole using only Newton's Laws, without the benefit of relativity and quantum theory.

Earlier this century, many of our leading physicists, such as Planck, Bohr, Heisenberg, Einstein, Pauli, Fermi, Dirac, Schroedinger and others encountered a similar kind of problem. Their experiments on fundamental particles such as electrons, protons and neutrons inevitably led to a number of paradoxes if they looked at them from a "classical" or "old physics" point of view. Attempts at building a theory from the old perspective met with contradictions such as the wave-particle duality, the uncertainty principle (the inability to specify both the position and motion of a particle), the distortions in their ability to measure mass, time and space, and, perhaps most disturbing, the literal, undeniable effect of the observer upon the thing observed.

As a result, the notion of a purely objective, dispassionate observer had to be discarded, much to the chagrin of many scientists, and a whole new physics had to be formulated to take into consideration these troubling effects. Out of all this was born quantum physics, whose main thrust became an attempt to look at the behavior of ensembles of particles by statistical means and to accept the paradoxes and dualities of particle activity as regrettable facts of life.

These are considered to be some of the main features of the "new physics." My previous book, *Exploring Inner and Outer Space*, and many other books explain the intricacies and implications of the new physics so I won't detail them here. The main point here is that it is often difficult to grasp the full impact of this science even though it has been with us for almost a century now. I myself have been a case in point: although I was teaching these theories in the physics department at Princeton University, presenting equations on the blackboard for students to memorize and regurgitate in a left-brained manner, I did not address what might truly be going on either physically or metaphysically.

One point we physicists seem to have missed, or brushed aside, is that paraphysics, in the form of the observer effect, is present to some degree in all experiments. The essence of the "new, new" physics goes even further: it is that our consciousness not only influences the behavior of small, subtle particles; it can also change the gross physical, chemical and morphological properties of everything with which we come into contact, whether it be an instrument, another person, a plant, a spoon, or our own bodies.

Thomaz Green Morton knows this well. Robert Jahn knows this well. Many of the pioneers of quantum mechanics know this well. Any psychic healer knows this well. And, as I described in *Exploring Inner and Outer Space*, the polygraph scientist Cleve Backster knows this well from his repeatable and replicated experiments in biocommunications and "cellular telepathy." What is at issue here is understanding how and to what degree this interactive effect operates. As we shall see later, this knowledge could crack open all the sciences and initiate a renaissance whose magnitude will surpass our wildest dreams.

The Dawn of New Theory

In *Margins of Reality* Jahn and Dunne take a crack at theory. They argue that the paraphysical phenomena that they measure have their counterparts in the basic principles of quantum mechanics. Using quantum theory as metaphor, Jahn and Dunne take us on a journey through the mechanics and mysteries of consciousness. They address the particle-wave duality (is light a particle or wave?) by positing that waves dominate when the "frequency" slows down and the waves become longer. We then have a situation in which the waves can penetrate and transform an object in a "free-flowing, holistic, generalized, aesthetic" manner. Particles dominate when the frequency speeds up, the waves become shorter, and the style becomes "more precise, reductionistic, specialized (and) analytical."

Physicist David Bohm and others have considered the possibility that the frequency and amplitude of the waves of consciousness may be primary information that can be Fourier-transformed (mathematically) into information about action in time and space that is familiar to us. This is similar to how a hologram works.

Putting mass, time and space into secondary roles in evaluating the mechanics of consciousness opens new possibilities of understanding not only pyschokinesis but also precognition and other distortions of "normal" temporal reality. These effects are well-known to parapsychologists, as well as to Jahn and Dunne, and pose difficult questions if we look at the force fields of influence only from the point of view of proximity in time and space, as in the other known forces of nature. Bohm calls the primary (frequency/amplitude) information the "enfolded, implicate order" and the secondary (time/space) information the "unfolded, explicate order."

No matter which model becomes selected, it appears as though human consciousness behaves like a vibrating energy field which interacts with the fields of objects within its sphere of attention. We will later see that these field-like influences do not necessarily respect time and space, so they do not behave like classical electromagnetic waves such as visible light, infrared waves and radio waves. Nor do they behave like any of the other three known forces of nature measured by physicists—gravity and the weak and strong nuclear forces. One almost certainly needs to postulate at least a fifth fundamental force in nature, and/or the existence of unexpected, nonphysical dimensions of activity, to explain the interaction of consciousness with the environment (see, for example pp. 70-76 of John White's *The Meeting of Science and Spirit* and Gary Zukav's *The Seat of the Soul*).

While many physicists now openly acknowledge the importance of consciousness in the equation, others ignore it entirely. The latter group optimistically seek a reconciliation of the four known forces of nature into a unified field theory which could provide a "final" answer about all the laws that govern the physical universe. The work of Stephen Hawking and other physicists in the development of string theory represents such attempts. However, though none of these theorists seem to consider the fifth force of consciousness, they have come up with six more dimensions beyond time and space—ones that are difficult to characterize physically yet seem to pop out of their mathematical equations into the real world.

So as we can see once again, theory is still seriously lagging behind observation. Therefore it is incumbent on scientists to develop new ways of approaching our inquiry. Otherwise we will continue to be stuck, arguing over stale issues (such as, "are these anomalous effects real?") and our leading university scientists will continue to steer clear of this embarrassing situation, continue to ridicule the entire field of inquiry and continue to hope the problem will go away.

And yet in this information age we live in, some of us have developed new tools to attack problems such as this. One such tool is systems engineering, and a bright, unusual person named Thomas Bearden has employed it to develop a very promising theory of the paranormal.

In the next chapter we take a high dive into the abstruse world of New Theory according to Bearden and others. If you find this information difficult to digest, please feel free to read it selectively or to skip the chapter entirely.

6

The Old New Physics and the New New Physics

Huntsville, Alabama, May 31, 1991

I was on stimulus overload while waiting to meet Tom Bearden at a gas station near his home. I had just completed a five day, five-city speaking tour through the Southeast, ending with a six-hour drive from Asheville, North Carolina to Huntsville, Alabama, where Tom lived. Just before that I had attended the tenth annual meeting of the Society for Scientific Exploration (SSE) in Charlottesville, Virginia (see Appendix).

The SSE had given me new experimental insights into the overwhelming evidence that our consciousness does influence the material world and that the effects do not necessarily respect time and space. After reading Tom Bearden's books and papers on his unified theory I felt I had to meet with him to discuss in greater depth the nature of the mysterious world of mind-over-matter, to get it from the horse's mouth, to see if this particular horse made sense or whether it was a theoretical dead-end.

Bearden is a systems engineer who works as a consultant to the aerospace industry. He retired from the Army as a lieutenant colonel and spent a good part of his military career in defense analysis, intelligence and rocketry. Huntsville is a high-tech, high-I.Q. city and claims to have more National Merit scholars per capita than any other city in America. But if this brain-dense environment produced an average I.Q. of 130, Bearden's seemed to soar above 200, perhaps in proportion to what I knew to be his large physical stature. He had the added attribute of having made it his personal passion for more than twenty years to understand the mysteries of paraphysics. While he was hard to follow at times, he seemed to be a pragmatic man with a mission and a desire to share his insights from a left-brained point of view—territory in which I felt very comfortable.

These were my reflections while waiting for him at the gas station. I also had some doubts and questions both about him and about myself. Could I keep up with his cerebrations about the nature of reality? How would I feel, as a former physics faculty member at elite Princeton University, knowing less physics than a systems engineer in the aerospace industry without a Ph.D.? How could I translate these important concepts into a form understandable to the lay reader? Could I find others to verify his work, based on careful study? Was Bearden the formulator of a workable theory of consciousness? These were my concerns as I once again turned on my car engine to run the air conditioner because I was perspiring heavily from the moist heat.

Bearden soon pulled up in an American car large enough for his frame. I got out of my car, he rolled down his window, we introduced ourselves and shook hands, and he told me to follow him to his house. I was amazed at his sheer physical size. He seemed to be about six-feet-four and weigh about 300 pounds. A greying, bearded, balding man of about sixty, he was wearing a coat and tie and was returning home from a day in the office.

Entering his house at the end of a street in a higher, cooler, wooded area, I walked into what would become a 24-hour marathon of brain food, interrupted by only six hours of sleep. His basement office was enormous, containing countless shelves of volumes on studies of the paranormal.

"I'm glad you're here, Brian," he said in a Southern

drawl. "Now is the time to get the word out. By the time you leave, you'll understand the mechanics of psychic phenomena, healing, gravitation, relativity, quantum theory, free energy, UFOs, crop circles, consciousness, and immortality."

In other words, the mechanics of just about everything!

Bearden introduced me to his wife, Doris, then poured two glasses of iced tea for us. He and I walked out to his back deck, where our discussion began (Figure 16). He came across as a sincere, personable, Southern country gentleman who happened to have an extraordinary flair for finding the keys to the universe.

For me to recount what he said in his own words would leave behind all but the most technical reader, and this is not a technical book. So I will attempt to capture the essence of his remarks, paraphrasing rather than quoting, as well as bringing in some ideas of others who have also been thinking about these issues for a good many years.

The Reframing of Physics

Bearden began with a reframing of the history of classical electromagnetics, now a well-known branch of physics. This history began more than a century ago when the brilliant Scottish physicist James Clerk Maxwell derived some precise mathematical expressions that described the puzzling behavior of electrical and magnetic force fields associated with light waves.

Any university student of physics or electrical engineering knows that Maxwell's equations are the fundamental basis for a wide range of experiments, instrumental designs, and theories of modern physics. In fact, electromagnetic phenomena underlie much of our present understanding of how the universe works. Various instruments can detect and describe the nature of the electromagnetic energies emanating from atoms, animals, people, planets, and galaxies.

These energies include cosmic rays, gamma rays, X-rays, ultraviolet waves, visible light, infrared waves, and radio waves.

Bearden explained to me that Maxwell had written his equations in a complex and obscure mathematical form known as quaternions, which were later misinterpreted by authoritative physicists who really didn't understand how quaternions worked. According to Bearden, the version of Maxwell's equations that emerged in the text books included the standard vector quantities of electrical and magnetic fields but left out an important and fundamental "scalar" term embedded in space-time through which the electromagnetic wave travels.

In order to proceed, a few definitions are in order here. Space and time (usually considered unitively and interactively as "space-time") are the familiar dimensions we deal with in ordinary physics and in everyday life. A vector is a numerical value plus a direction (i.e., "go fifty miles per hour, due east"). A scalar, however, is a quantity without direction and has only a numerical value; in other words, the vector sum of a scalar is zero. A scalar can be seen as a point in space-time; although it has a value, it is essentially hidden from sight and can appear not to mean anything (i.e., "go fifty miles per hour in no direction!"). But that can be very deceptive.

Bearden pointed out that two papers by the British mathematician E.F. Whittaker in 1903 and 1904 provided the needed clarification for formulating a complete theory that included the scalar term as well as the vector terms. The scalar term omitted in physics texts turns out to be a form of gravitation rather than traditional electromagnetics. This conjecture would send most physicists into immediate reaction, because gravity has always been considered to be separate from electromagnetics.

But the type of gravitation Bearden was talking about is not the relatively weak and diffuse pull of the Earth, Moon,

Sun, planets and stars. It is instead a concept known as "electrogravitation," caused by a very small-scale curvature of space-time which produces enormous, but randomly fluctuating, force fields. (Similar concepts are incorporated into David Bohm's "hidden" or "implicate" order, which he has written about in several popular books.)

The concept of curvature of small-scale space-time naturally follows from Einstein's theory of general relativity and quantum theory. But because these forces operate in such a tiny space and cannot be measured by existing instruments, they have been of no use to us. "That's why," Bearden asserted, "we have gotten away with ignoring the scalar term in Maxwell's equations for so long."

According to this theory, if we could "order" the local force field we could tap enormous quantities of energy. The ordering process could be achieved either through using the human mind or by building an electrical system that could produce the right interactive force fields. Bearden believes this can be done by "scalar interferometers" that can focus the energy on any point in space-time from a distance. One form of interferometer would consist of two separate generating coils through which an electrical current would be run. Another would be the human mind through which the current of consciousness is run.

Bearden suggested that herein lies the secret of psychic phenomena and healing, as well as the potential for "free energy" to solve the world's energy crisis. I would have easily dismissed him as another crank were it not for the work of dozens of other outstanding investigators with diverse philosophical outlooks who have come to similar conclusions. As we shall see, some of these individuals have actually developed "free energy" devices to prove their point. And of course we have the mystery of mind-over-matter that makes equally little sense in terms of conventional physics.

Another way of looking at this situation comes from the theoretical physicist Hal Puthoff, who wondered how electrons orbiting their nuclei don't collapse into the nucleus as they radiate their energy away. Following the work of others, he then deduced that there exists a vast source of energy in the "vacuum"—the fabric of space itself— to sustain the electrons. David Bohm concluded that one cubic centimeter of empty space contained the energetic equivalent of several hydrogen bombs, but in potential form. It is not released into space-time; it is "implicate," or still enfolded, rather than "explicate" or unfolded.

Perhaps the first scientist to look at these possibilities was Nikola Tesla, the inventor of alternating current. Much has been said and written about this man—his genius, his extraordinary experiments, and the challenges he faced in dealing with the existing economic and political structures. Much of Tesla's work on free energy remains enshrouded in mystery, apparently because of his awareness of its enormous economic and military potential. Bearden and others "outside the pale" have carried that concern to this very day. At this point we appear to have experienced nearly a century of underground experiments, both within and outside official government circles, to produce free energy—as well as psychic weapons of mass destruction.

Bearden continues to see the related physics in terms of a unified field theory— something that that has for so long eluded scientists. But in order to accept all this, says Bearden, we would need to revise three of our most powerful theories of contemporary physics: classical electromagnetics (we must add the scalar term); general relativity (contrary to earlier thinking, we must look at the curvature of small-scale space-time which produces "gravitational fields," which, when ordered, can provide enormous amounts of energy); and quantum mechanics (no longer are systems of particles thought to have to behave randomly—under certain conditions we can engineer order into such a system and extract vast amounts of energy). This last point needs some more explanation, and we can bring in some of the ideas of others besides Bearden to clarify it.

I find it interesting that many of the leading theorists in this field are systems engineers; in other words, engineers trained to be generalists. Most mainstream scientists are, by contrast, specialists. The latter are usually ill-equipped to address the complex relationships that exist among such diverse subjects as classical electromagnetics, relativity, quantum mechanics, statistical mechanics, thermodynamics, and general systems theory, to say nothing of acknowledging the possible reality of mind-over-matter and free energy. All this seems to be too much for anyone besides the generalist to handle.

Moray King's Theories

Moray B. King is a systems engineer in Provo, Utah, who has spent the last fifteen years thinking about how a free-energy device might work from a theoretical perspective. While his language and presentation are quite different from Bearden's, the basic message is the same: it is likely that we can "order the vacuum" and extract vast amounts of energy from it. Unlike Bearden, he bases his theory entirely on broadly-accepted concepts in modern physics and engineering. This is a less revolutionary approach that may be more palatable to the traditional scientific mind.

I met King at the annual conference of the U.S. Psychotronics Association in Dayton, Ohio, in July 1991 (see Appendix). A thoughtful and gentle man of about forty, his quest began when, as a graduate student in systems engineering at the University of Pennsylvania, he wondered if antigravity were really possible. His interest derived from reliable eyewitness reports of UFO phenomena. He was impressed with the quality of these observations, as presented by reputable scientists such as J. Allen Hynek, Jacques Vallee, Richard Haines, and others. He was especially intrigued with the erratic, bizarre, non-Newtonian motion of the reported UFOs.

King acknowledged that Einstein's theory of general relativity involved the production of gravitation by the curvature of space-time attributable to an energy source E in which $E=mc^2$, where m is the mass (or equivalent mass to the energy) of the attracting object and c is the speed of light (about 300,000 kilometers per second).

The important point here is that Einstein and other physicists felt that these relativistic effects could only be produced near massive stars and black holes; small-scale effects would not be important. One argument was that one would need a nearby mass of more than a million tons to levitate a UFO against the Earth's gravity—unless empty space or the air were somehow filled with energy that was so abundant that it had that mass-equivalent. Yet Puthoff argued that such energy must be present to maintain electrons in orbit about their nuclei. The thought seemed inconceivable to most mainstream scientists who generally express disdain toward the "impossible" claims of outsiders about anti-gravity. But Puthoff was not the only physicist who argued that such an energy may be present.

King quickly learned that many other maverick researchers also recognized such a possibility. Textbooks on gravity, and the pioneering work on relating gravitation to quantum theory by my former Princeton colleague and Nobel-laureate John Wheeler, suggested to King that there exists an all-pervading energy embedded in the fabric of space consisting of fluctuations of electricity. This is called "zero-point energy." The term "zero point" derives from the fact that in a condition of absolute zero, when all molecular motion ceases, the energy measured by conventional means appears to be zero. In reality, however, if it could be tapped it would be absolutely enormous.

Wheeler, King, Bohm, Puthoff and others have now recognized this possiblity. "Quantum mechanics," wrote King in his book *Tapping the Zero Point Energy*, "showed that this energy was constantly interacting with matter and the elementary particles in what is called vacuum polarization." If only a small amount could be made coherent, King thought, then we would have free energy and anti-gravity.

When King quizzed his university professors on whether zero-point, or free, energy could be tapped, most replied that they didn't know this energy had even been hypothesized. I certainly hadn't known about it as a physics faculty member at Princeton. Even those who did know about it said it would not be possible to tap it because of the grim reality of increasing randomness and disorder of physical systems, according to the well-established concept of entropy (the second law of thermodynamics). Traditional physics clearly did not permit the ordering of any of these systems. Instead, it insists, we must wait for the energy of universe to slowly wind down.

Enter chaos theory. In 1977, Ilya Prigogine won the Nobel Prize in chemistry for his work on identifying the conditions under which a physical system could evolve from chaos to order. Much to his surprise, Moray King found that these conditions could be applied to the kinds of systems containing random fluctuations of zero-point energy propounded by Wheeler and others, and could therefore theoretically be tapped. His conclusions were the same as Tom Bearden's: It is possible, in principle, to create the conditions of anti-gravity and free energy production by performing some smart engineering.

The basic conditions for ordering such a chaotic system, according to chaos theory, are that it be far from equilibrium (that is, it must be turbulent or unstable), and that energy be run through the system to maintain this instability. The energy required needn't be that much; just enough to keep things stimulated. Starting with Nikola Tesla, a number of inventors have taken a crack at producing zero-point energy, and some appear to have been successful. One such system proposed by King uses a hot gas containing charged particles, called a plasma. When enough energy is added, the system can order itself into a ring known as a plasma vortex, a phenomenon similar to ball lightning.

King has suggested that by running a current through carefully-designed coils, it may be possible to extract zero-point energy from such a vortex. He cited three devices already developed that appear to produce more energy than they use, based on similar principles.

Back to Tom Bearden

As I sat on his back deck listening to his story unfold, Bearden's commanding presence and engaging intellect made a powerful impression on me. Often I would paraphrase his remarks back to him to find out whether I had correctly understood his theories. I found that to understand them I had to suspend or unlearn much of the physics I had always thought to be the final word. Another reminder that in science there is no such thing as a final word!

Bearden was a man on a mission, with a curious mixture of right-wing politics (fear of a Soviet plot to subjugate Americans by means of psychotronic weapons), patriotism (working with the defense industry on state-of-the-art weapons systems), and suspicion of top-secret U.S. government and elite private-sector activity (citing sinister attempts to cover up, discredit and threaten the work and the lives of those involved in open research on UFOs, free energy and psychic phenomena). I felt my own mission here was to set aside his politics (even though I acknowledged that some of his fears could conceivably turn out to be true) and to understand what kind of new scientific framework would be needed to facilitate high-quality research on these concepts, from both a theoretical and experimental perspective.

King, Bearden, Wheeler, Bohm, Puthoff and others have devoted years of their lives to discovering and then articulating their theories, and we are all struggling to develop a common language to describe what is going on. These research findings are often expressed in language hard for even a technical person such as myself to understand, and it has taken repeated exposures for me to gradually embrace these theories.

But in spite of the lack of appropriate attention paid to these developments by the mainstream media, government, industry and the scientific community, it seems clear to me that we do have here a body of theory, much of it based on accepted principles of modern physics, that can make the notions of free energy and psychic phenomena very palatable— to those individuals willing to take the time to examine it.

Where Do We Go From Here?

A natural question to ask is, why don't we use this energy now rather than continue to rely on oil, coal and nuclear power, all of which are expensive and environmentally unsound?

That is a difficult question to answer simply. For one thing, the discoverers and inventors usually tend to be loners. For another, they are often intimidated by, or impatient with, the pressures, procedures and expenses of obtaining patents, and by a chronic lack of cooperation by established industry. Even though we are talking about potential multi-trillion dollar business, the news is not good, looked at from the point of view of entrenched interests in oil and other energy sources. Also, in most cases we are not yet at the point of having commercially-viable models, although that day is closer than we might imagine. Meanwhile, the work goes on under secrecy and in fear of co-option.

I should mention here that Bearden and others have described to me active efforts to supress the dissemination of free-energy technology, UFO-derived information and technology, and other such "weird stuff." This seems to me to be a field ripe for careful investigative journalism, but no mainstream journalist has, to my knowledge, as yet stepped forward (perhaps the territory is too fraught with career-threatening implications). What we are talking about here is nothing less than solving the energy crisis along with much of the environmental crisis— no small prize! Meanwhile, denial and fear seem to carry the day.

I should also add that I have been delighted to discover a robust underground network of inventors of other kinds of free energy devices, some of whom appear to have open patents for all to see and who have built demonstration models. For example, Moray King has presented an operating device invented by William Hyde. Bearden discussed a device invented by Howard Johnson. I have received similar information from Reed Magnetic Motor, Inc. All these devices use rotors to manipulate the electrical or magnetic fields in such a way that they seem to extract more energy than is put into the system. King and I have theorized that so-called cold fusion may, in reality, also be a form of free-energy extraction.

Some Further Reflections and Implications

Bearden and King have spent a long time synthesizing for themselves, and for those in the technical community who might be interested, a model for ordering the incredibly abundant, randomly fluctuating energies that pervade every point of the universe.

The key seems to be to apply a slight, continuous energy force to a given point in a turbulent space-time matrix. Under certain conditions, according to the widely-accepted chaos theory, that force can trigger the release of a newly coherent energy in which small-scale space-time is curved. The trigger can be either a carefully constructed device, a plasma vortex, or it could be our own consciousness.

One important aspect of this theory is that effects are not limited to local regions, as is normally dictated by conventional theories of gravitation and electromagnetics. "The holistic paradigm," wrote King, "has the zero-point energy as the source that maintains the elementary particles and, therefore, all matter. It has recently been shown to be the basis for the stability of the hydrogen atom. Bohm shows that it is the basis of the implicate order from which arises the explicate phenomena of matter, energy, time and space. Bohm's implicate order contains a quantum potential that results in nonlocal correlations across space-time (as well as perhaps across the multiple, parallel universes of Everett). These nonlocal linkages result in a holistic description of our universe as a hologram—where the whole view is implicitly embedded in every section of the hologram. Note that Wheeler's hyperspatial wormholes likewise generate a nonlocal connectivity. The zero-point energy constitutes the first substrate of organization, and allows phenomena to be linked nonlocally through a higher dimensional space."

Bearden essentially agrees with King's assessment, but emphasizes different aspects of the question and speaks a somewhat different language: "We can either add or subtract enormous amounts of energy at any given point in the world by pulsing scalar waves from interferometers (devices that can focus the energies of two coherent scalar waves at a given point, at a distance). The Soviets have already developed powerful weapons and weather-control devices based on these concepts, but we can also use this technology to produce free energy and to heal virtually any disease. The choice is ours to apply these technologies for our common good."

My 24 hours with Tom Bearden were hypercerebral, pushing the edge of the envelope of my comprehension of the nature of reality, which already had been turning out to be stranger than fiction. Bearden's reframing of physics had alarmed me at first blush because in order to comprehend it I had to unlearn so much. In the end I was convinced he was not just another crank or egomaniac, and that he sincerely wanted to share with me the fruits of his years of labor. We said goodbye and I welcomed some time alone on my flight home to assimilate the enormous implications of his ideas.

Could it be that free-energy devices, spoon-bending, UFOs, crop circles, the outrageous antics of Thomaz, and the carefully controlled experiments by Jahn and Dunne, all have something in common?

I believe so. We are dealing with a new, new physics. This physics goes far beyond the widely accepted "new physics" of quantum mechanics and relativity. But before we can jump to such a conclusion, we need to look at the experiments, explore the theories further, and gain more insights on the mechanics of consciousness, healing, telepathy, free energy, and their holistic, nonlocal nature. Experiments and theory expressed again and again in different ways by different people comprise Jahn's and Dunne's concept of the "two-step" between theoretical and empirical science that marks true progress.

Herein lies the key to the new scientific paradigm. From all this, in my opinion, will quickly evolve transcendent technologies that will transform the world. The bigger box we are building for the role of science in the tapestry of reality is beginning to take shape.

Casino Consciousness

Las Vegas, Nevada, November 27, 1991

My professional reason for visiting this mecca of gambling and debauchery was to host a television show on UFO research. But I also thought that as long as I was there I'd try my luck at the slot machines and blackjack tables.

I had just spent a month of virtual seclusion in the Oregon mountains, so on this first night out I felt optimistic and welcomed the excitement of the casino environment.

As I walked by the slot machines and blackjack tables I stopped at each one I felt good about and dangled a small pendulum. Why a pendulum? It's a device employed for dowsing, or divining. I was using it to check on whether or not it would be worth playing that particular machine or table. I tried to do it discreetly, but nevertheless some people did look at me strangely.

Dowsing is an ancient technique used by traditional peoples—and by increasing numbers of new scientists, particularly in the former Soviet Union where the practice is known as "biolocation." In the best-known traditional type of dowsing, the practitioner uses a forked stick or other natural object to locate or "divine" water, minerals, lost objects, etc.

Techniques vary, but all seem to depend upon some as yet poorly-understood principles of interaction among the dowser's mind, his or her subtle body-energies, and the intended target situation. In some types of dowsing the device or tool seems to respond directly and physically to the presence of the target, while in other cases the operator's subtle muscular movements, amplified by means of a pendulum or other ballistic device, indicate the presence of the target or of a negative or affirmative condition.

Although dowsing is never 100% consistent, in the right hands it can give yes-or-no answers to questions with surprising reliability. For example, "Is this where I can find water?" or "Will I make money from this machine if I play it ten times?" Some modern dowsers have had excellent track records in finding missing persons, stolen goods and so forth, by means of "map dowsing," in which a map of the target region may substitute for the actual, physical location.

I often use simple, L-shaped wire dowsing rods quite effectively to map the changes in the "auras" of people to whom loving energy has been directed. I am learning to use the more convenient (and pocketable) pendulum, which seems to work most effectively when I ask specific questions, when my own consciousness is clear (that is, when I'm feeling "centered"), and, most importantly, when I am not emotionally invested in the outcome.

That first night in Las Vegas I picked some good slots and tables and won some money. A feeling of high spirits and laughter pervaded the tables. It was a good night for me.

The next day I worked hard on the video project and it turned out well, but by evening I was worn out. Nevertheless I wanted to "celebrate" its success at the casino. Again I used the pendulum, but it didn't seem to work and I began to lose money. Several times I stubbornly hung out at a particular machine or table, gritting my teeth and hoping my luck would turn around. The more I lost the more I feared losing, and the more I feared the more I continued to lose. Was I creating a self-fulfilling prophecy?

Apparently so. Obviously I was in what we all know as a losing streak. I had visions of entering a "black hole," a low state of being, a helpless "Fear and Loathing in Lost Wages!"

The next day, when I was feeling better, I had two big wins at the slots just minutes apart and won some of that money back. But that night and the next I sank back into a black hopelessness as I experienced another long losing streak. The more emotional I became, the more useless was the pendulum.

Most of us can identify with my episode. We all have winning and losing streaks that often seem to be related to our attitudes. Professional skeptics might argue that these are merely the vicissitudes of random events or, in this case, rigging by the house. The new scientist asks instead, "could there be an observer-object interaction here that significantly affects the outcome?"

I would tend to answer in the affirmative because these effects appear to be much greater than the one part in several thousand which Jahn and Dunne have seen from operators in the laboratory. But because the criteria of the old, rigorous science demand the compilation of large statistical samples of data, with the final result averaged over time as the primary information worth looking at scientifically, we would naturally expect to come up with marginal results. Also, an "experimenter" in Las Vegas is emotionally invested in the outcome and is therefore not as dispassionate as he is in a relatively antiseptic laboratory!

But what about the fine structure of the data? Are these winning and losing streaks an important factor in the mind-matter interaction? If so, what is going on here? Are there even more subtle influences on these patterns? Can we apply filters to the data to extract larger effects?

Recent experimental results answer these questions with a strong "yes," and provide considerable additional insight into what actually goes on during human-object interactions.

The Spindrift Experiments

Over the past sixteen years, a group of Christian Scientists known as Spindrift, Inc., have performed a number of experiments that are beginning to shed light on these important questions. Unbound by the rigid experimental protocols that normally constrain parapsychologists, the Spindrift investigators have used double-blind and other control techniques to get positive results using the healing power of prayer on germinating seeds, on the growth rate of yeast cells and on the progress of heart patients. They have also found large effects in the interactions of operators with random event generators and with cards concealed inside envelopes. The most important key to their success appears to be the manner in which the data are analyzed.

Their first such experiment was so simple it could be tried by virtually anyone; you can easily try it for yourself. Called the VIUR test (visual image, unconscious response) it employs a deck of 24 cards inside envelopes. The cards are either black or red, twelve of each. The operator is asked to shuffle the envelopes at least six times, then go through them one at a time and "guess" which of the two colors is inside. He or she would record each guess on a data sheet until the trial ends. The experiment is repeated several times.

The normal way to analyze the data is to record the cumulative "hits" and "misses." With a total of 240 calls, we would expect about 120 hits with an expected error spread of plus or minus a few on either side of 120. Indeed, that was the case for most operators, and is consistent with the very small effects measured by Jahn and Dunne, who require much larger samples of data.

We are learning, however, that unconscious assumptions are often incorporated into methods of analysis. Such assumptions can then hamper the discovery of valuable information contained in the data.

This became apparent when the Spindrift scientists

developed a different method of analysis. What they did was to separate the calls into two sequences—one corresponding to the black image calls and the other to the red image calls. They then looked at both sequences for successive pairs of hits or misses to see if they were identical (hit-hit or miss-miss) or were different (hit-miss or miss-hit). A redefined "hit" became those pairs that are dissimilar (hit-miss or miss-hit).

Spindrift found that, analyzed this way, the data yielded extraordinarily strong mind-over-matter effects. For eight operators, they produced a statistically significant average of 52.5% "hits," or about one part in twenty. The effect got even larger (about one part in eight) when the data were selected for alternating pairs from the original list of hits.

What does all this mean? The data suggest that we are influenced by an ordering mechanism that wants to alternate hits and misses. In other words, we all seem to want to create a result that evens out any tendency to bias something in one direction or another. We tend to want to restore order to the kingdom, to mitigate our own volitional psychic powers. Stated differently, if we don't apply a filter to the data that sidesteps the unconscious saboteur, the volitional psychic effects decline to Jahn's one part in 5000.

From a collection of experiments, the Spindrift team has come up with a theory that four forces of consciousness are interacting in parapsychological experiments and in spiritual healing:

- The first force is the volitionally psychic force used to guess or influence the right cards, slot machines, blackjack tables or binary numbers produced by random event generators. This is a goal-directed force which can bend spoons and seems to come from the solar plexus. The motto behind this perceptive force is *"my* will be done."

- The second force is a defense mechanism, an inner voice that tells us not to achieve the goal. It basically tells us (unconsciously) we shouldn't exercise our psychic powers, because the unconscious mind somehow feels threatened by using these powers.

    - The third force is an ordering force, one which leaves it to a higher power ("God," or "the superconscious") to create the most harmonious pattern "for the highest good." The motto for this force is *"thy* will be done."

    - The fourth force is, again, the unconscious mind wanting to sabotage the achievement of the third force.

    The Spindrift experiments suggest that these hypothesized four forces appear to be at cross-purposes, like four teams of horses trying to pull a load in different directions at once. The two sabotaging defense mechanisms are the most powerful throughout, keeping the other two forces largely in check. These are the forces that keep us from being totally psychic and/or totally spiritual. The volitional force is the weakest, measured to be only one part in thousands for most operators, as compiled by the Princeton group. The ordering force appears to be significant, creating up to 10% effects depending on the operator and how the data are analyzed. This last force is what Spindrift is studying extensively in examining the power of prayer and of healing.

    Ironically, both the Christian Science Church and the parapsychological community frown on Spindrift. The church's dogma keeps it from embracing the work, and so the investigators have chosen to remain anonymous. The parapsychologists have also questioned Spindrift's credentials, methodology, and the unorthodox nature of their inquiry.

    Clearly, we have here a situation of a group of new scientists who have fallen outside two boxes! They have wanted to share their results with these two groups, their own church and their community of scientific peers, but neither group has wanted any part of the results.

    This is almost a textbook case of old science and religion clashing with new science. In order to make significant progress, we find once again the need to expand the box to allow for new approaches and possibilities while keeping the scientific method intact. In this case Spindrift came up with some fresh thinking about how data could be analyzed, and the results have enormous implications. Such an experimental approach would never have occurred to mainstream parapsychology.

The Importance of Replication

Have these important experiments been replicated by others? They have, by Dean Radin, a well-respected American parapsychologist at GTE Laboratories in Chantilly, Virginia. Radin admitted to being skeptical at first because he knew well the common pitfalls in experiments conducted by those outside the profession. Any number of errors could creep into the experimental procedure that would keep the skeptics of the world going for years.

The danger of being ridiculed is a strong inhibitor to innovative approaches in the much-battered community of parapsychology. Of all groups, professional parapsychologists most scrupulously obey the skeptics' adage, "Extraordinary claims require extraordinary proof."

And yet, to Radin, the experiment seemed so colossally simple and the results so clear and repeatable, with so little effort required to accumulate significant statistics, that he felt it was worth a try. He thought he could easily find fault in Spindrift's approach or analysis that would lay to rest their claims of having discovered large effects.

Along with his collaborator, George Cross, at GTE Laboratories, Radin ran the experiment and, as expected, found no strong effects when the data were analyzed in the conventional manner. Then they applied what they thought was Spindrift's method of filtering out pairs of hits and misses. Still no significant results.

But just as they were about to give up, Radin decided to call Spindrift, and during the call discovered that he had used the wrong data filtering criteria. Apparently, rereading the self-published Spindrift papers several times was not enough for the replicators to understand exactly how the data were to be analyzed for the ordering effect. When Radin and Cross reran the analysis the right way, they came up with a strong positive replication of the Spindrift experiment.

From all this comes another lesson for New Science. In order to come up with the needed fresh approaches to questions, the traditionalists and the innovators must first understand each other's languages. In this case, Spindrift did not understand the language of parapsychology and vice-versa. (Lamentably, this kind of misunderstanding often occurs between specialists in different disciplines of a more conventional nature.)

The simplicity of this experiment was fortuitous. But it is to Radin's credit that he persevered, in the face of potential criticism from colleagues, to decipher the works of the innovators and, perhaps even more importantly, to communicate with them and directly engage them in dialogue. As a result Radin and Cross were able to give these important results the blessings of rigorous scientific replication, and were able to present the strongest mind-over-matter effects with ordinary people ever measured under controlled conditions.

Replicating the Backster Effect

A similar set of circumstances surrounds the work of Cleve Backster. For more than two decades this world-renowned polygraph (lie detector) expert has picked up electrical responses in living matter that turn out to be precisely time-correlated with the emotions of a person who is connected or associated (no matter how distantly in space) with the living matter under test. This could be white blood cells donated by the subject, or even a favorite houseplant.

Backster has successfully empoyed a variety of other organisms such as brine shrimp and live yogurt cultures in a range of similar polygraph experiments. As I reported in *Exploring Inner and Outer Space,* I have done several experiments with Backster and have satisfied myself that the effects are quite real.

Backster's results have been published, yet the parapsychological community still refuses to give him the time of day. In a 1988 report to the National Research Council, the Parapsychological Association dismissed Backster's work as "not representative of parapsychological work in general...[the research] does, however, illustrate research in which a plausible, empirically-grounded alternative explanation does exist and a negative conclusion is justified." (Translation: Because there *might be* an alternative explanation, therefore there *is* an alternative explanation. Hence, applying the scientific method, in the form of actual replication of Backster's experiments, is unnecessary. Our minds are made up.)

Theodore Rockwell of Bethesda, Maryland, is a solidly credentialed nuclear engineer who spent much of his career in civil service working in a technical capacity to develop nuclear submarines for the legendary Hyman Rickover. He is an experienced, productive, well-grounded and highly regarded scientist. "Having declared Backster's work outside its domain," Rockwell has written, "the Parapsychological Association then proceeds to render expert judgment on it." He has pointed out that such a sweeping statement invalidating twenty years of careful research without so much as informing Backster or attempting to replicate his work is an example of scientific dishonesty.

Fortunately, the Backster effect has been replicated, by John Alexander of the Los Alamos National Laboratories. Several similar experiments are now appearing to validate this concept of biocommunication: In addition to other work by Alexander, there are the well-controlled experiments of the respected German biophysicist Fritz Popp and investigations by the American biologist Michael Pleass.

Already many investigators have independently found that biological systems can interact with one another in ways that stimulate each other's electrical properties, and that these effects are what quantum physicists term "nonlocal." In other words, they do not necessarily respect distance. These findings add an important piece to the puzzle of the New Science although they fall outside any of the boxes represented by well-established, traditional disciplines.

Today an interesting pattern is beginning to develop, one in which some of our greatest breakthroughs may be realized: innovators outside the mainstream are making claims that are then verified by those within the mainstream who have the courage to try replicating the work. This kind of "intercredential" teamwork among strange bedfellows could well be the wave of the future.

Radin and Cross could hardly conceal their excitement about their successful replication of Spindrift's startling findings, albeit in the somewhat stilted language of a parapsychology journal. "We find the present result intriguing," they recently reported, "for two main reasons: First, the experiment was accidentally conducted double blind, and second, after the data was collected it took several false starts before we understood the proper method of analysis."

"In these previous and the present studies," they concluded, "the general idea has been to take data which shows chance results for direct hits, then reorganize or reanalyze that data in an attempt to find predicted time-based or pattern-based structures. In Spindrift's case, their four mental forces are conceived as working together to arrange the calling sequence in such a way that it appears that nothing of interest has occurred. To achieve this splendid illusion, a fairly sophisticated simultaneous decision process must actually be unfolding."

Using the Forces of Consciousness

Radin and Cross then went on to suggest a model that incorporates the interaction of the four forces that adds up to a call decision. Because some of the effects are as great as one part in five, one could propose designing an electronic feedback system that could be responsive to one or more of these forces. Its responses could then be incorporated into a biofeedback system that might help make us more aware of how our consciousness interacts with the environment.

The "ordering" force is the one that appears to be most easily isolated from the others during the course of a sequence of calls of binary numbers from an electronic feedback device. It may also be the force most important to us, since it transcends our goal-oriented psychic will as well as our unconscious desire to sabotage the results.

We already know that through subjective means (for example, self-hypnosis and meditation) we can "trick" the saboteur and sidestep the volition of our egos. This allows us to tap into a higher order which, in its wisdom, tends to restore peace to a chaotic situation and to heal the imbalances of the environment. But here we may be talking about something more accessible to ordinary people: nothing less than a healing machine; an example of transcendent techology for the next millennium.

So what happened to me in Las Vegas? I suspect that my fear carried the day there, interspersed with some interludes of confidence. If I had had a little device with me that prewarned me to balance pairs, perhaps I could have adjusted my bets and my ventures. Calling red or black in roulette might be a first application for such a discerning device, designed to reveal significant patterns hidden from our conscious awareness.

Gambling houses, beware!

8

Ashes and Diamonds

Whitefield, outside of Bangalore, India, Jan.22, 1992

My anxiety level rose again when our bus jerked to a halt to the bells of a railroad crossing. How long would this one take? One minute? Five? Ten? Twenty? Thirty? One hour?

For two weeks we had traveled extensively in and around Delhi, Agra, Jaipur and Ahmadabad, dazzled by what we had seen of the unending, unrestrained drama in the streets. Nothing was hidden from view, including people relieving themselves, having cathartic fits in the Hindu temples, sacrificing goats, and being cremated in the open. We had seen a lack of sanitation and a ravaged environment; yet we also saw that life went on for the millions of Indians we had seen by the roadside, for the most part unaware of our own provincial, antiseptic ways. We had experienced all of the above and much more in this puzzling kaleidoscope of a nation-subcontinent called India.

We had also experienced an unbelievable variety of inconvenient delays. We waited eight hours at the Udaipur airport because the next airport on our puddle-jump to Bombay was closed because the prime minister was there. As a result we missed an important meeting in Bombay with one of the leaders of the Indian environmental movement.

But this delay at the railroad crossing felt more significant. We were about to see the renowned Indian swami,

Satya Sai Baba. Our group of fourteen miracle-seekers had flown to Bangalore for two days to see if we could at least catch a glimpse of this amazing man, considered by many to be a saint and perhaps the avatar of our time (as Jesus Christ had been to the Western world two thousand years ago).

Regardless of what one might believe about what Sai Baba represents in the religious world, at the very least, he does attract crowds. Currently, they are the largest for any living spiritual or political leader in India. He has the reputation for being a "living national monument" who appears to materialize things in front of thousands of people during his nearly daily morning and afternoon darshans (audiences).

The previous day we had taken the three-and-one-half-hour pilgrimage by bus from Bangalore to Sai Baba's principal ashram in the village of Puttaparti, hoping to catch his afternoon darshan. We had missed the great guru by hours, as he had left Puttaparti for his smaller ashram at Whitefield just outside Bangalore—where we were staying! His movements are sometimes hard to predict; all we knew was that he was about to travel to Bombay and that we might therefore miss him entirely. Our primary goal was to catch him before he left.

Even though we had not seen him in Puttaparti, we did have a revealing glimpse of the place where he and his followers live. It was once a tiny village—the village where he grew up. There, over fifty years ago, at the age of fourteen, he had been bitten by a scorpion. This event marked the acquisition of his special powers. Puttaparti is now a clean, thriving metropolis in the middle of a hilly southern Indian desert, complete with hospitals, universities, museums, libraries, and shops, all geared toward the teachings of this one miracle-worker. The abundance of large colorful statues of religious leaders, elephants, and pastel buildings suggested to me the name "Saibabaland" for this complex.

The full story of Sai Baba's life is presented in several biographies, of which the most scientific is *'Miracles are My Visiting Cards,'* written in 1987 by the University of Iceland parapsychologist Erlendur Heraldsson. This book is a detailed compilation of firsthand observations and interviews with dozens of scientists, devotees, and former devotees attesting to the authenticity of most if not all of Sai Baba's miracles. More on that later.

Meanwhile we were stuck at the railroad crossing less than a half-mile short of the Whitefield ashram, and we might already have been late for the swami's (brief) morning darshan. This was the last day we could get to see him, as we were scheduled to fly to Calcutta the next morning. We had worked hard to get this close to the great master. Why was the train so slow? Was this a test? I reflected that one needs the patience of a saint in India.

Two nights earlier, when we arrived in Bangalore, we had networked ourselves into a meeting with a Mr. B.N. Srikantaiya, one of Sai Baba's principal devotees. We spent three hours in the home of this gentle, intelligent, giving and humble man, an engineer who had gradually begun to practice meditation and other spiritual activities with Sai Baba. We chanted mantras with him and expressed our desire to meet with the great swami. Srikantaiya made no promises, and said it was far better to have no expectations, but rather to allow the events of our visit to unfold in inner ways. In other words, any anxiety about seeing the outer man Sai Baba would probably only stand in the way of getting to see him. Contact on the inner planes, on the other hand, could lead us directly to the source on the outer planes. Our group began to accept this philosophy, feeling gratitude for the opportunity to participate in an uncertain adventure where actually seeing Sai Baba in person became a secondary goal.

The primary goal was to let go of worry and preconceptions, to become spiritually attuned, and to enjoy the process. As we began to "get" this, Srikantaiya seemed to pick up on this acceptance, and so began to go out of his way to create for us a potential audience with Sai Baba. To the hard-driven, successful westerner it seems ironic that the real secret to staying in the process is simply to let go.

As I sat on the bus waiting for a second train to pass, I thought about Srikantaiya's words. I also reflected on those times when success came my way as soon as I was willing to release intense expectations—whether it was trying out for NASA's astronaut program (I was accepted at the moment when I let go of my doubts) or looking for phenomena in the English crop circles as we shall see in Chapter 9.

I also observed Srikantaiya sitting next to me in calm anticipation of the events of the day. At last the second train passed by, the barricade went up, and we were on our way. As we arrived at the gate of the ashram we were relieved to find that people were still gathering for the morning darshan. My anxiety had been totally unnecessary! We were also greeted by one of Sai Baba's top staff people, who arranged with Srikantaiya for the men in our group to have front row seats for watching the swami walk by. The women were less lucky, sitting about four rows back on the other side of the aisle (the Hindu tradition is that men and women are separated when in an audience of a swami).

Srikantaiya motioned for me to sit next to him. It was at this point that I wondered whether the word had gotten out that I was a former astronaut, and whether this credential alone provided us with our privileged place in the front row! Such considerations would surely be secondary to a man of such spiritual accomplishment. Or would they?

After some minutes of wondering about this as well as meditating about the events to come, I realized that the audience of about two thousand had begun to hush. Heads began to crane toward a gate next to the mandir, which was Sai Baba's private quarters in Whitefield. The gate was some two hundred yards away, and some uniformed guards began to shuffle in the area near the gate.

Suddenly and softly the swami appeared. Wearing a saffron-colored robe with long sleeves, he almost seemed to be walking on air as he slowly strutted to a small grouping of people near the gate. He waved his hands and rose petals appeared to drop from his hands onto those of the others. His Afro-style hair formed a haloed silhouette against the morning sun, adding a dramatic touch to his appearance (Figure 19).

It was an awesome moment, not in terms of idolatry or devotion (although plenty of that seemed to be going on as well), but in terms of encountering something of the transcendental, of temporarily leaving behind the mundane.

As he further approached us, now about 150 feet away, I took some telephoto pictures of him: he flicked his hand and moments later a puff of vibuti (sacred ash) appeared and floated down to the outreached cupped hands of some women in the front row (Figure 20). The women were giving him sealed envelopes which he seemed to bless, and after some accumulation, would give them to one of his assistants walking near him. I later found out that these sealed letters were primarily requests for healing, which he often fulfills simply by placing his hands on the envelope.

As he headed closer to our group I put down my camera. Now only about fifty feet away, he was now almost upon the men in our group. Srikantaiya then coached us to ask Sai Baba for a private interview. Now he was right in front of us, and one of us asked for an interview. He smiled and produced vibuti that streamed out to our cupped hands from his right hand about a foot above ours, one dose at a time. I smudged some of it onto my forehead and tasted it (quite an achievement for a Western non-devotee such as myself, as traditional vibuti is literally an ash resulting from burning cow dung!).

Srikantaiya then introduced me to Sai Baba in Hindi. The swami walked back to some of the other group members and asked them how many of there were in the group. When the trip co-leader Kevin Stevens said fourteen, Sai Baba said "go inside." We paused in some disbelief, although Srikantaiya was getting the drift and he was aglow. Sai Baba repeated, "Well, go inside." He then wandered on.

We got up, motioned to the women in our group across the way to come with us. It all felt strange yet exhilarating to walk toward Sai Baba's private mandir in front of thousands of people, many of whom would give anything for the privilege. Even Srikantaiya hadn't had this opportunity for years.

Sai Baba's assistants led us through the gate and lined us up, once again with the men and women separated, outside the mandir. As we waited, the skeptic in me reflected that there could be a slight chance he could have brought the rose petals and vibuti out from his sleeves through sleight of hand. I then realized how unlikely that was: when he was standing just above me, the quantity of vibuti he produced was substantial. Its very fine consistency would have made it virtually impossible to manage in that way. This observation I later found to be validated and well-documented by thousands of others. But it was best for me to see for myself first.

The Private Interview with Sai Baba

As we waited outside the mandir, Sai Baba reappeared, walking in front of his college students and talking with the principal. He then walked up a ramp from where the students were and entered into the mandir, a simple round building with several doors around the periphery suggesting a landed UFO. He soon appeared in the door nearest us, walked halfway down the ramp, and motioned for us to come inside. I felt the same kind of awe and wonder one might experience in being invited into the craft of a superior, benevolent alien race.

He greeted us one at a time. When he saw me he said he knew all about me and we would be talking inside. His

English was understandable for the most part, although his use of it was unique to our ears. He was a small man, only about five feet tall. His eyes were dark, deep set, and seeming to shine.

As soon as we got inside I was surprised to find that Sai Baba would be seeing us alone; no staff members or bodyguards were present. The room was simple, with two doors, one to the outside and one to the inside, small barred windows, and a cushioned floor seat for Sai Baba.

The rest of us sat on the floor either lotus-style or with our legs straight out. As soon as he came into the room, Sai Baba grabbed me by the hand and sat me down right next to him. He smiled at me as if I were an old friend. Then, when it was becoming obvious I was sitting on the women's side, he shuffled me back over to the men's side.

He directed his first remark toward me. "I came to you in a dream last night. Do you remember?" I said I didn't know for sure but was trying to recall. Underneath all this I felt coming from him a sense of a deep personal interest in me, which opened what was to become just about the most extraordinary ninety minutes of my life. Was it just my credentials or the real me he was after? This was a question I have often asked.

Kevin Stevens said that we were a group of seekers travelling India to observe miracle workers such as himself. Sai Baba answered, "Seekers of what? For who? You don't know who I am. I'm not my body. See my hand. There's nothing here now."

He then turned toward the women and in a clockwise motion with his right hand showered their outstretched hands with vibuti. From his vantage point, Kevin had seen a supply of vibuti between the swami's thumb and forefinger. Sleight of hand, or a materialization? Our consensus was that the latter was true.

Our group was joined by a Mr. Karachi from Pakistan, who was a Vice President of the World Bank. Sai Baba told him he knew everything about him and would give him a symbol of his devotion. He then waved his hand and produced a ring for him. He lectured him that money wasn't everything. He also told us that we had "too much baggage"; I wondered whether he meant strictly the physical kind.

Just before producing the ring for our Pakistani friend, Sai Baba was seen by one member of our group to look and reach underneath his cushion as if he were checking for his watch. Sleight of hand? In this instance it may have been possible, though we kept in mind his extensively-documented reputation for being authentic.

He then talked about spiritual matters. He and Srikantaiya exchanged some words about the mantras we had chanted, and Sai Baba referred to these as "diamonds" inside of us. Sitting about eight feet across from me, the swami then directed his attention toward me, waving his right hand clockwise. As he did this there appeared in his hand a beautiful gold ring set with a large diamond-like jewel (Figure 21). He handed it over to me saying "this is yours, your diamond, your essence."

I tried it first on my left hand ring finger and he said, "no, your right hand!" and so I worked it over my large right hand ring finger knuckle. He talked some more about our jewel within, and then asked if my ring fit, or whether it might be too big. I noticed that it flopped around my finger and that it was a bit too large. So he asked for it back (Figure 22), I handed it to him, he blew on it twice, and he handed it back to me. Although now it was even more difficult to wrestle over my knuckle I did manage to get it on again and the size was perfect.

Between the two sizings, one member of our group had noticed him once again looking and reaching down to his cushion. Sleight of hand? Possible, of course, but somehow the new ring size was ideal.

Later that afternoon a woman in our group reported seeing what looked like a glittering gold column extending from about two feet below his hand just prior to producing a ring, as if it were being teleported from below. Sai Baba himself says the process of materializing things is simple: he just thinks about it and it is there.

I was overwhelmed by his gift to me. He asked, "Are you happy?" and I exclaimed "Yes!" He was to ask the same question of me several times later during the interview.

Sai Baba continued to talk in metaphors and riddles. He held out a kerchief and asked, what is this made of, cloth, cotton or thread? Cotton is what you think you are, he said, thread is what others think of you, and cloth is what you really are. He equated the conscious mind to the ego, conscience to mind, and consciousness to spirit. Some of these concepts didn't seem clear when he first broached them in his unique style of English.

I mentioned to him the New Science movement in the United States and wondered how he felt about combining science and spirit. "What is science?" he asked in a riddle. "This, this, that, that, this and that. There is no new science. Science is like a C." He put up his hand in the form of the letter C.

"See, see," he continued, then putting his other hand in an opposite-C, completing a circle. "Now we have another C, the C of spirit making it all whole, an O."

Then he talked about the unconscious (sleeping), subconscious (dreaming), conscious (awake state), and superconscious (the next higher level of spirit which he says he inhabits). He continued with a riddle saying that the number one is greater than nine because $9 = 1 + 1 + 1$, etc.

He talked about there being nothing on the Moon, it's only reflected light, that we had to bring everything with us to survive. He directed some of these remarks to me personally, almost as if he thought I had been there. Perhaps he had thought I was astronaut Ed Mitchell, who had actually been to the Moon and was interested in many of the same things I was.

We asked Sai Baba about the foul air of India and the environmental destruction of the Earth. He criticized governments, gurus, and knowledge from books, saying what was important to solve these problems was a change in our consciousness through our experience.

Sai Baba directed some personal remarks toward some of us, often stating explicitly that he already knew all about us, our past, present and future. Yet he occasionally missed the mark, sometimes asking leading questions like an amateur parlor psychic. But he seemed to be correct on the most important matters.

For example, he appears to have played a significant part in healing one of our group of a serious chronic condition. Mary Stuntz had been a victim of L-Tryptophan poisoning in 1989, which destroyed many of her leg muscles and nerves. Mary had written Sai Baba about these problems and requested a healing in the form of a letter she placed in front of her during our audience.

She reported his looking into her "very essence" during the course of the interview. As we later adjourned into a circular interior room, Sai Baba asked Mary, "Your legs, you have pain in your legs." She answered yes. "They are like arthritis pains," he continued. She said yes. He said, "It will be all right." Mary internally registered "I know that my legs will be restored to perfect health" and viewed this as her own part of the healing process.

That night Mary's legs felt hot, very hot to my touch. Soon feeling began to return. Just two weeks later, Mary reported that her legs had been restored about "65% to 80%" back to normal.

Sai Baba said to another one of our group, "You have bad blood." (She was diabetic.) With a wave of his hand he produced some rock candy (!) for her to help her healing. She has since reported significant progress and has cut her insulin intake in half.

He also correctly perceived one of our group member's husband's medical problems and his inclination to worry, as well as another woman's divorce and fights with her ex-husband.

All four of these women remain convinced of Sai Baba's psychic and healing powers. Yet another woman in our group was very disappointed that he didn't respond to some of our requests, that he seemed to ignore the women a lot, and that he flaunted his ego so much. At the same time she admitted that some of these perceptions may be a reflection of her own psychological projections and insecurities.

Debriefing and Aftermath

In many ways our privileged private audience with Sai Baba was as much a mirror of ourselves as it was an objective view of a miracle-maker. During the lengthy debriefing session we had later that night, the fourteen of us shared fourteen different opinions of the nature of the encounter. Some saw Sai Baba as another Jesus, others as an egomaniac; some saw him as an ordinary but talented man having a bad day, others saw him as the greatest healer alive; some saw him as having greeted us as if we were heads of state because of my astronaut's credentials, others saw the visit as a total action of spirit, of divine grace. It could have been all of the above. We humans can be very complex.

After a photo session in the interior of Sai Baba's mandir (Figure 23), we parted, but not before the swami grabbed me by the arm and said that two women wish to marry me (This was hardly far-fetched!) but that I shouldn't do so. He said that being married would mean four minutes per day of happiness and 23 hours and 56 minutes of compromise and grief. I still ponder his remarks and realize that they could be interpreted in various ways. In the end only I can resolve that question, yet I can well imagine what an effect such advice would have on many a devotee.

We boarded the bus and it stopped again at the same railroad crossing— this time mysteriously, because no train was coming. We noticed that one of Sai Baba's most trusted assistants was waving down the bus and asking to see Srikantaiya. After a short conference between the two of them, Srikantaiya got back on the bus all aglow. He said that just after our private audience Sai Baba had requested that *I* give a major speech to his college students that very afternoon after his darshan. He would share the podium with me. I was overwhelmed and speechless.

Kevin Stevens answered "yes" for me.

And so that afternoon we returned to Whitefield. I was nervous, but also very much on a "high." What a privilege it was to be asked to speak. I didn't have my usual slides to show, so it would all need to come from my heart, cutting across large cultural distances. But I felt ready.

Sai Baba greeted three of us—all men from our group—just outside his mandir near the gate, within eyeshot of the thousands gathered for his afternoon darshan (See Figure 24). I gave him an autographed copy of my book and he gave me some pictures of us from the morning.

"When are you leaving?" he asked.

"Tomorrow morning, to Calcutta", I said.

After some pleasant small talk, he asked again, "Are you going to Calcutta? Tomorrow morning?"

"Yes," I answered.

He even asked again, almost as if he didn't believe it. Then he said we would be very busy, so it would be best to return some time in the future. This conveniently relived me of having to give my speech!

He personally invited me to return to see him in Puttaparti, asked me when that might be possible and I said January of 1993. He said good, that I should plan on spending at least a week with him (this I plan to do, in connection with leading another group there in 1993). Then, still holding my book in his hands, he walked toward his waiting flock (Figure 25).

Interestingly, we did not go to Calcutta the next morning because the flight was delayed until that evening. Perhaps in questioning me repeatedly about our intended departure, Sai Baba was trying to tell us something.

The Sai Baba Phenomenon and the New Science

In *'Miracles are My Visiting Cards,'* Erlendur Heraldsson well documents over forty years of paranormal phenomena produced by Sai Baba. Still, in some ways my visit with this great man was not as thoroughly convincing as my experience of the materializations, phenomena and transformations of Thomaz Green Morton because there was slight residual doubt about what might be up his sleeves or under his cushions.

Heraldsson bases his work on eyewitness accounts on an almost daily basis of materializations, sometimes of specific items on demand, under a variety of circumstances. Several eminent scientists and even a professional magician-skeptic have vouched for the preponderance of Sai Baba's miracles as being authentic and not sleight of hand. He has also been observed repeatedly to bilocate, disappear, reappear, produce vibuti remotely, etc.

And so once again we are forced outside the box. Even though their personalities are very different (except for the fact that both of their egos are very strong), Sai Baba and Thomaz dramatically and repeatedly seem to disobey the rules and regulations of Western materialistic science (and if you don't think the phenomena they produce are real, I strongly suggest that you visit either or both of them to see for yourself). In order embrace these phenomena we will need a science that goes beyond materialism. Otherwise we will be like the blind leading the blind in circles.

9

Agriglyphs and Agraffiti

Alton Barnes, England, August 1, 1991

The mature wheat rippled in a gentle breeze and the day was cloudy. Forty of us stepped off the bus parked in a farmer's driveway just a few miles from the famed stone megalith at Avebury. I was the scientist-leader of this, the first organized tour of Americans destined to experience and study the mysterious crop formations in England (Figure 26).

The "crop circle" phenomenon, in one form or another, has been with us for decades, perhaps for centuries. In its simplest form it consists of swirled and flattened areas in fields of wheat and other crops that appear mysteriously and then disappear when the fields are harvested. At first the phenomenon was assumed to be the product of random whirlwinds, or of pranksters.

Almost all of these formations, however, possess characteristics that cannot be produced by natural meteorological phenomena, or by mere mechanical action such as that of people wielding ropes and boards. For one thing, the affected plants remain undamaged and continue to grow, albeit horizontally. In some mysterious manner individual plant stalks (even the more brittle varieties) are bent, rather than broken, as if some kind of high-energy force field has caused them to "melt" and bend over at their bases, and then to resolidify. Another extraordinary aspect of the crop circles is the presence of anomalous energetic and acoustic phenomena, some of which have been recorded by scientists and by broadcast television crews.

Mysterious circles have been detected all over the world, on every continent except Antarctica. But the most abundant and elaborate formations have cropped up in southern England, especially during the past several years. Starting in 1989, many formations appeared there that were far more complex than circles. These included features and details such as rings, triangles and dumbbells, as well as forms resembling musical notes and weather-map flags. Some of the more whimsical formations have been strongly suggestive of insects, fish and dolphins.

Many explanatory theories have been put forth in terms of poorly-understood or newly-hypothesized natural phenomena, such as "plasma vortices." But one by one these have been eliminated for a variety of reasons, not the least of which is the many levels of intelligent design exhibited by these formations. Not only are the overall shapes intriguing, but the fine-scale structure, visible only upon close and careful examination, seems to contain a bafflingly complex, elegant combination of charcteristics. For example, the plant stalks are often bent over in distinct layers, each interwoven with the next. A single crop formation may include various combinations of clockwise and counterclockwise swirls, as well as radial, circumferential and other patterns. The long axes of many geometric crop formations lie precisely parallel to some important nearby natural or man-made feature such as a road or field boundary, or to the "tramlines" made by tractors through the crops.

Some of the more recent crop formations seem to resemble ancient Scottish petroglyphs, others Vedic (ancient Hindu) symbols, yet others Hopi symbols, and so forth. But despite many interesting attempts, nobody has definitively "decoded" the meaning of these features. There have been as many theories as people proposing them.

A recent event led to a tremendous misunderstanding

about these "agriglyphs" (a term coined by ufologist Bruce Maccabee). In September 1991 the mass media dismissed their authenticity, suggesting that every one of them—thousands, in fact, going back many years—had been hoaxed by two English gentlemen in their sixties named Doug Bower and Dave Chorley. "Doug and Dave" had obliged the media by staging a demonstration in which they produced crude facsimiles of small crop circles using planks and chains. The media asked no critical or probing questions and enthusiastically reported Doug and Dave's "confession" at face value, implicitly endorsing it and leaving the public with the misperception that the mystery had been solved—even though the two fellows later recanted many of their sweeping claims.

In early September, on the CBS evening news, Dan Rather certified that Doug and Dave had done it all. This was apparently the final word for many intelligent but misinformed observers, some of whom undoubtedly slept better that night believing that the problem had been safely disposed of.

Meanwhile, investigators have linked the initial release of Doug and Dave's story to a shadowy London "press agency" whose identity is now untraceable. And increasing scientific scrutiny continues to reveal in the crop circles something that mainstream institutions still do not want to hear: that we have on our hands a major mystery that clearly challenges us to step outside the box of the old world-view and of the old science.

Back to Alton Barnes

Our own tour of the English agriglyphs took place before the media blitz that alleged they were hoaxes, so there was a certain purity of inquiry to our quest, free of mass misperception and the limelight of ridicule. We also had an extraordinary team of guides: Richard Andrews and

Busty Taylor.

Richard Andrews is a gentleman-farmer, a native of the crop-circle regions of Britain and an expert dowser (Figure 27). One of his interests has been the mapping of the so-called "ley lines"—lines of subtle energy, detectable by dowsing, that criss-cross the countryside forming a web or grid in which the sacred sites, such as those at Stonehenge, Avebury and Silbury, seem to form major "nodal points." Through the years, Andrews has noted that the agriglyphs tend to occur where these ley lines intersect, as if they were deliberately sited there. Many of the more spectacular crop formations have tended to appear within a very few miles (sometimes within a stone's throw) of major sacred sites.

Busty Taylor is an intrepid aviator who gets out there and photographs and maps these formations, often before anyone else arrives at the sites on foot. He also carries in his car a forty-foot pole with a camera mount on top, designed for ground-based "aerial" photos of the pictograms.

Together, Andrews and Taylor took us out to the main sites at prime times of agriglyph creation—in late July and early August. Their assistance turned out to be essential, as many of the farmers do not appreciate random trespassing of their fields. Richard and Busty would meticulously obtain prior permission to enter, and the admission charge was usually an affordable one pound per person.

And so for our first adventure we walked silently along a foot-wide tramline (tractor-rut) up a hill to a formation that had appeared just two nights previous. As we walked into the formation we all experienced a feeling of reverence, of which it was difficult to make a left-brained analysis. This was no ordinary tour.

Both Richard and I had first shared with the group the importance of allowing our own inner intuitions to unfold as we walked through the formations, being open to whatever we felt rather than distracting ourselves with idle chatter about the many technical details. Sizes, shapes, times of formation, etc. could be addressed later. Here I reflected once again that the New Science will need to incorporate both inner and outer methods of approach, each in its own turn.

As we began to step into our first agriglyph, one woman in our group began to experience an emotional catharsis and burst into tears. Many others remained silent and isolated in meditation. I myself felt awe and an inner peace. My L-shaped metal dowsing rods registered a strong presence of energy, and in some places spun around like the blades of a helicopter—something I had not previously experienced (Figure 28).

Ironically, accompanying our group was a reporter from the Daily Mail (a widely-read British newspaper) who complained that she felt nothing happening. She later wrote a scathing article on how flaky this whole business of our tour was and how unscientific this whole exploration was. Could this have been a case of seeing only what she expected to see, of sour-grapes envy of our joyful feelings, or perhaps of pressure from her editors to conform to the mainstream view? We'll probably never know.

Although we could not view this formation from above, it was clearly among the more unusual agriglyphs. Later aerial photographs revealed it as an extraordinary fish-like shape, fins and all, with rings around the head and tail (Figure 29). In places the wheat had been flattened in several successive layers in various directions, suggesting a complex sequence of formation. The boundaries of the "fish," in chest-high wheat stalks, were distinct and precise, and the whole thing smacked of elegant art. Of all the radical explanations we had heard proposed, the notion that two gentleman from a pub in their sixties could have created all this with chains and planks seemed by far the most radical of all.

Gazing down the hill from the "fish" we saw another smaller formation that was oval-shaped with circles at each end. Richard told us this feature had formed just the night before and therefore was free of human trampling and of possible other forms of pollution and degradation. So we eagerly walked down the hill to this newly formed agriglyph. A helicopter began to circle overhead, disturbing the peace for a few minutes. Later we learned this was the well-known author and expert on ley lines John Michell producing a show for the BBC. We were not alone!

The second formation felt to me even more awesome in its purity. Many of us felt an inner peace which inspired us to meditate and to chant together. Again, the stalks were bent at their bases to a horizontal direction, stayed alive, and were swirled in an elegant way as if some artistically inspired energy force field were at work. I reflected here that seeing books and pictures about this phenomenon is not enough. There is no substitute for actually being in its presence and experiencing its many dimensions first hand.

That afternoon we went out to another, older feature that had formed about two weeks earlier near Barbury Castle a few miles north of Avebury. While this pictogram had already been degraded by human activity, its geometry was intact. Photographs taken by Busty Taylor revealed it to be complex and elegant, structured around a huge equilateral triangle with spirals and circles at its apexes and with several other connecting lines as well, spanning an area about the size of two football fields. The alignments were accurate to within inches over the entire formation, which would require surveying instruments of great accuracy and impeccable care in their execution. Another case of "chains and planks," no doubt!

The Genesis of Crop Formations

Careful scientific studies by a number of responsible British scientists seem to reveal that at least some of the crop formations were formed in a matter of seconds.

Laboratory studies reported recently by the American investigator Michael Chorost show a definite change in the chemical and crystalline structure of the wheat stalks at the points where they are bent.

What is going on here? My own theory is that the energy or force field impinging on the plant stalks is created by some form of intelligence aware of a "technology" related to that involved in spoon-bending. Recall that when a spoon is bent, the metal literally softens during a "time window" that lasts for only a few seconds.

Even if we knew how to employ conventional electromagnetic forces, such as microwaves, to produce such precisely-modulated phenomena, the hardware involved would be relatively bulky and would have to be placed awkwardly close to the "target" crops. Any boom, blimp or other device large or steady enough to hold such equipment would be easily seen by the local populace or detected by the heavily instrumented all-night vigils that have been organized by scientists and the British army.

On the other hand, nontraditional electromagnetics, such as those of Nikola Tesla or Tom Bearden, might possibly explain the mechanics of formation. As in spoon-bending, some kind of force field could be produced by the interaction of some form of consciousness with the crop fields, perhaps with the help of certain consciousness-responsive equipment.

The next question is, what kind of intelligence might be behind all this? Why hasn't the real cosmic artist stepped forward? One would think that the real source (or sorcerers) would have revealed themselves by now.

I recall here an important common denominator of my experience of Thomaz and of Sai Baba: these miracle-workers both seem to have strong human egos and a great appetite for exhibiting their talents. The crop-circle artists, by contrast, have kept their indentity secret for many years and seem to be operating under some long-term plan for the gradual revelation of their "message." I am normally cautious about giving specific interpretations of unusual phenomena, but I am willing to go out on a limb about this one. It seems unlikely to me that the responsible agent is human.

So this leaves some kind of nonhuman intelligence, some kind of cosmic artist or artists who use the wheat fields as a canvas, not only to portray shapes and symbols, but also textures, levels and myriad details within those shapes and symbols. In addition, the 1991 formations revealed "signatures," indicated by two small circles a few feet in diameter near the formation, one swirled clockwise and the other swirled counterclockwise.

As a final gesture, our group spent two hours with Richard Andrews one night in a wheat field near the city of Winchester—an area where agriglyph formation had been prolific. During this time I led a meditation in which I suggested we might see one of these features actually forming. When a mist came across the field, I led the group out of meditation and also released any expectation of our seeing anything paranormal. I also expressed gratitude for the opportunity of being together in this beautiful setting in nature. Suddenly, about ten to twenty feet behind me, a small red ball of light started to dance rapidly and playfully around the wheat stalks, and just as suddenly disappeared. Because my back was to it I missed it, but over half of the fifty people present did claim to have seen it.

The behavior of that object was consistent with that of small silver discs that have been videotaped in daylight dancing through some of the English wheat fields. (Various other luminous UFO phenomena have often been observed in close proximity to fresh crop formations. One of these has been documented on a British army videotape.)

Our experience in the crop circles seems once again to reaffirm the importance of releasing anxious expectations and of maintaining a sense of gratitude. As I have attempted to convey in this book, this approach is a prerequisite for success in the New Science.

Richard Andrews and other investigators are beginning to predict, by means of dowsing, where future crop formations might occur. Other researchers, such as Colin Andrews (no relation to Richard) claim to have "wished"— without serious expectation—for certain kinds of forms, only to have them appear in novel crop circles within days. This is the beginning of an interactive approach that parallels a current trend in UFO research known as "close encounters of the fifth kind" or "human-initiated contact." We shall see in the next chapter that this approach is beginning to pay off.

Close Encounters
of the Fifth Kind

St. Malo Retreat Center, Allenspark, Colorado, May 20, 1992

The thunderclouds began to unleash rain and lightning just as Maury Albertson, Paul Von Ward, Scott Jones and I pulled into the circular driveway of this large, modern building nestled in a stunning valley surrounded by the fir trees of the high Rockies.

There was still a spring chill in the air at the rarefied altitude of 9500 feet. Having just flown in from San Diego, I panted and puffed as I ascended the stairs with my luggage laden with books and papers. The St. Malo retreat, a lovely facility containing many luxurious meeting rooms as well as living suites, was recently built by the Catholic Archdiocese of Denver. For the next two days it was to become the unlikely site for a first-of-its-kind UFO research think-tank, which would be followed by a three-day public symposium at a Denver airport hotel. I was one of the organizers, and felt a special calling to jump in and participate.

The idea was the brainchild of Maury Albertson, co-founder, with myself, of the International Association for New Science. Albertson, a professor of civil engineering, was a founder of the Peace Corps and former vice president of research at Colorado State University. His success in attracting 80% of the University's research money had given CSU the reputation of being the primary recipient of university research funds west of the Mississippi. Though he is in his seventies, his efforts to open minds to new paradigms continue to be tireless and effective.

Albertson's main premise for this kind of think-tank event is that if the top people in a field can be brought together in a relaxed, personally-revealing context, they will synergistically create high-quality funding proposals that can support every reasonable research project. Having successfully achieved this goal in funding other kinds of projects, Albertson aleady knew that funding agencies appreciated this kind of cooperation and mutual support among the principal scientists in a given field.

We were well aware of the risks involved in attempting to do this in the field of UFO-related research. Perhaps no field of inquiry has been more riddled with controversy, saddled by governmental and mainstream-scientific ridicule, attacked by unenlightened skeptics and crucified by tabloid headlines. And yet, thanks to the energetic efforts of IANS administrators Carol Singer and Bob Siblerud, twenty-two of the twenty-five top people on our worldwide list of UFO researchers accepted our invitation and were soon to show up at the St. Malo Center (Figure 30). Some participants flew in from as far as England and Belgium.

My first job was to greet these eminent people and to suggest ways in which we might tackle our chores in such a short period of time. That evening, my friend Paul Von Ward, CEO of the Washington-based nonprofit company The Delphi International Group, facilitated a remarkable, personally-revealing introductory process that broke the ice among many veteran UFO researchers as well as among the rest of us. It went well into the night and set the stage for a surprisingly harmonious weekend.

The next day we divided into groups according to discipline: Stanton Friedman, Jerome Clark, Walt Andrus, Don Schmitt, Bruce Maccabee and Patrick Ferryn focused their efforts on solidifying the credibility of historical events. Of particular interest was the 1947 Roswell, New Mexico case in which many dozens of eyewitnesses have testified that at least one flying disc did crash there and that the government has been sequestering and covering up the evidence. Other high-credibility, multiply-witnessed and well-documented events include the numerous sightings at Gulf Breeze, Florida, the recent military and civilian sightings throughout Belgium and the many mysterious cattle mutilation cases with which a UFO connection has become increasingly evident.

A second group looked at the UFO abduction phenomenon: Rima Laibow, Albert Stubblebine, Edith Fiore, Leo Sprinkle, James Harder, Ellen Crystall, and Richard Haines—all professionals well known in the field—took part.

The third group was devoted to crop circle phenomena research and included English investigators Richard Andrews and Busty Taylor as well as American researchers Michael Chorost, Paul Von Ward and myself.

A fourth group considered the social implications of these phenomena, and included John Salter, Scott Jones and Maury Albertson.

The largest of the groups was chaired by Dr. Steven Greer, a pioneer in human-initiated UFO contact, also known as "close encounters of the fifth kind," or "CE-5" in ufologists' jargon. Greer, a brilliant, well-spoken emergency-room physician from North Carolina, claims to have begun interactive contact with UFOs in Belgium and Florida early in 1992. Videotapes of the latter encounter demonstrated how Greer's patterned flashing of searchlights elicited identical responses from the enigmatic hovering objects. Clearly this was the "hottest" new item of UFO research to appear on the horizon.

Other think-tank participants included television producer Linda Moulton Howe, Mars photoimaging expert Mark Carlotto, writer-reseacher Dan Drasin, media producer Elia Wise, and psychologist-ufologist Michael Brein.

After two days of work and socializing (and getting an occasional glimpse of Mt. Meeker, a 14,000-foot mass of snow and rock towering over the retreat center), we forged the drafts of twenty-five proposals, soon to be ready to send to potential funding sources.

One thing we all understood was how underfunded UFO research efforts have been. Most of the best research to date has been voluntary, with occasional small stipends for specific work. Interestingly, those familiar with government contract work, such as Scott Jones, Mark Carlotto, Richard Haines and Maury Albertson, initially came in with substantially higher funding requests than did the seasoned UFO researchers, who have somehow managed to prevail in spite of zero institutional funding. This was a meeting of two very different populations: those whose livelihood still depends on governmental or institutional funding sources and those who have braved it as entrepreneurs. I myself shifted from the former to the latter group about five years ago.

The total requested funding for all projects was in the millions of dollars spread over the next few years. It was a small sum by today's standards, yet the return could be enormous: we would be able to upgrade the historical record (Roswell, of course, plus existing files of tens of thousands of sightings and other encounters); be responsive to new phenomena on a global basis (dispatch top researchers to active sites); do the necessary and timely physical studies of crop formations; examine more openly the implications and applications of UFO phenomena and derivative sciences and technologies; begin earnest team efforts to initiate contact with UFOs; and hold major international conferences on UFO research that would coordinate these efforts and provide forums for disseminating the results. In this field the stakes are high, but so may be the rewards as we gradually lift the veil of denial and ridicule.

**An Institute for Research in Extraterrestrial
and UFO Studies**

One very important outcome of the think-tank was
the concept of a UFO research institute that would ensure
high quality research. This received virtually unanimous support from the disparate free-thinkers present at the retreat.
Such an institute would serve as a professional home for
UFO researchers, help provide them with research funds
and publish a journal including responsible peer review of
their work.

While the idea sounded good to all of us, some legitimate concern arose about the hazards of elitism and fav-
oritism. The last thing we wanted to do was to establish
another National Academy of Sciences, whose new members are elected by existing members in a sometimes very
political process in which the main criteria include the Ph.D.
degree, apropriate affiliations, professorships, publications...
and, unofficially, often the right personal connections. We
also recalled that one of the Academy's committees had sum-
marily dismissed the scientific value of 138 years of research
in parapsychology!

At first we had envisioned that such an institute be a
traditional academy with rigorously selective entrance criteria, because lines did need to be drawn at some point to
keep the quality of the research high. Then Paul Von Ward
made the enticing and bold suggestion that the institutes we
form be self-selecting. At first I thought the idea ill-advised,
as I visualized charlatans and loudmouths taking over the
organization and sinking it. Then Maury Albertson again
came forth with a pearl of wisdom: he pointed out that this
approach had already worked for our New Science Forums.
He felt that interviewing candidates periodically, and then
letting them decide for themselves whether they want to be
members, should be sufficient to make the selection. In most
cases, he felt, those individuals who were not qualified would

be able to see for themselves they would not be in the right place.

The first conferences Albertson hosted on Paranormal Research in 1988 and 1989 had attracted many well-meaning metaphysically-inclined folks who set up booths where they sold crystals, read palms, etc. To the casual observer these conferences may have more resembled a psychic fair or expo rather than a scientific meeting, which, however, was also going on behind the scenes in concurrent sessions. But in later conferences the fringe elements began mysteriously to vanish. The meetings in 1990 and 1991 were solidly scientific forums that attracted a good balance of professionally-qualified scientists and well-informed laypeople.

In my view the principle of self-selection may become a cornerstone of New Science. In other words, individuals would themselves choose to participate, and would not be judged a-priori by others as being out of line. While the new scientist would conduct his or her work with a high degree of ethics (for example, no weapons research or harmful bio-engineering) and approach it using the most appropriate available scientific methods and tools, the fact of member-ship and participation would not be an issue except where ethical codes or reasonable groundrules were violated. Insti-tutes based on these practices would provide valuable oppor-tunities for many individuals who may not otherwise qualify for participation on the research teams of, say, an Institute of Noetic Sciences or a Society for Scientific Exploration (see Appendix).

Asheville, North Carolina, March 3, 1992

Perhaps no individual better epitomizes the spirit of New Science than Steven Greer, M.D., of Asheville. Fol-lowing our private think-tank retreat, we held a public sym-posium in Denver, where, in a banquet speech, Greer shook the audience with his accounts of human-initiated interactions with UFOs over Belgium and Florida.

One of my first close meetings with Greer was in March 1992 when I arrived at his house, an elegantly-remodeled Tudor mansion adjacent to the Biltmore estate. It was a pleasant, balmy evening but I was exhausted from a speaking tour and from nausea I had experienced since the night before. In appropriately doctorly style, Greer gave me three Tylenols and sent me upstairs to my room for a refreshing nap. This was followed by a healthy dinner with him, his wife Emily, his four young daughters, and a recently-acquired golden retriever who occasionally made her presence known in the dining room. After dinner I spoke at the local Unity Church to a group of over 200. Then Steve and I stayed up late talking about some of the remarkable work he has done and about his plans for the future.

Still in his thirties, Greer is a true Renaissance man. His interests include practicing the Baha'i faith , being with his family, remodeling his home, riding his bike along the Blue Ridge Parkway, working as chief physician in a local hospital emergency room and acting as founder-director of the Center for the Study of Extraterrestrial Intelligence (CSETI). This last activity has been taking more and more of his time, and has involved travels to UFO-active areas, a growing number of public speaking engagements, and the leading of training sessons in contact and commnications protocols. Fortunately, he seems to need little sleep.

Greer is rapidly becoming somewhat of a folk hero due to his apparent success in initiating two-way communications with UFOs by means of light-signaling and through a disciplined telepathic technique he calls "coherent thought sequencing." He had achieved his first success in Belgium just three weeks prior to my visit.

The Encounters in Belgium

Ufologists are familiar with the remarkable wave of

sightings that took place over much of Belgium in 1990 and 1991. At that time, large triangular objects were seen in the sky by tens of thousands of Belgians in over 3000 documented sightings. Even the Belgian Air Force cooperated with UFO investigators by openly confirming the extraordinary speeds, sizes and shapes of these UFOs reported through visual sightings and documented by photographs and radar measurements. (A far cry from the attitudes of the tight-lipped US Air Force and intelligence community.) Americans saw this story reported briefly in installments of the nationally-broadcast television programs *Hard Copy* and *Unsolved Mysteries*.

On February 9, 1992, Greer and his team members, using his coherent thought sequencing technique, appear to have summoned six to eight illuminated objects that danced in the sky above a distant ridge somewhere in North Carolina. "The lights formed circles," said Greer, "and for a short time winked back and forth to us in answer to our lights."

·"The next night at a different site," he continued, "there was a very low cloud cover.... Around midnight, one of us saw a large light in the sky through an opening in the clouds...Shortly thereafter, we heard an intensely deep rumbling vibratory noise from the sky.... We felt convinced that a very large craft was above us, but could not come below the very low clouds. We felt the noise was their way of acknowledging us."

The Gulf Breeze Encounters

But Greer's greatest coup appears to have happened on March 14, 1992 in Gulf Breeze, a suburb of Pensacola, located on Florida's Gulf Coast, whose inhabitants had been reporting and documenting the presence of myserious aerial objects for several years. Greer put his results this way:

"A direct and unequivocal first degree CE-5 occurred involving a CSETI training group and five Extraterrestrial Spacecraft (ETS). This event was witnessed by over 50 people from two locations and resulted in audio tapes, video tapes and photographs of the objects.

"The training team was successful in vectoring into the area five ETS which interacted with us via light signals for over ten minutes. At one point, when we directed them to form a triangle by first drawing a triangle in the sky with a light, three spacecraft formed a perfect equilateral triangle in response...the salient points include:

"Congruency of time and place—The spacecraft appeared approximately 30 minutes after completing on-site coherent thought sequencing, while we were using lights and the CSETI tones (recordings of sounds picked up from a crop circle and UFO in Canada). They appeared very near us: a few thousand feet away—not several miles.

"Confirmed 'lock-on' with consciousness—Several members of the team observed via mental remote viewing, several spacecraft en route to our location with occupants aboard.

"Confirmed 'lock-on' with light signals and thought— Once the craft were visibly seen, we were able to signal in a reproducible fashion which resulted in their signaling back with a definite winking of their lights. Additionally, a link-up with thought and consciousness occurred indicating a high level of joy and excitement from the occupants, as well as our group.

"Once we signaled the craft to move towards us, they did so, flashing in sequence to our light."

While Greer's communications are unquestionably rudimentary, they constitute an important first step. If they can be further developed and substantiated we will obviously have made a major breakthrough in UFO research.

If this occurs, our traditonal scenario of detecting feeble radio signals from distant solar systems (and our simplistic notions such as that of a UFO landing on the White House lawn and its occupants asking to be taken to our leader!) may have to be reconsidered. To me, it is a more appeal-

ing scenario that a motivated team of independent citizens be
the cosmic icebreakers, and perhaps our first legitimate
ambassadors to the Universe. We must not forget that such
citizen diplomacy played a similar key role in helping to radically transform the West's relations with the former Soviet
Union, often circumventing the formidable obstacles erected
by governments.

Greer's work has barely begun and has not been without controversy. To further verify it and to help it evolve, we
will need trained "quick response teams" employing various kinds of techniques, protocols, and instrumentation, and
ready to travel anywhere in the world where UFO activity
is strong. Needless to say, such activities are not for the weak
of heart, given the thousands of reports by solid citizens
about their UFO-related abductions. Nonetheless I would
be delighted to participate in such a group; after all, can we
generalize about the motives of extraterrestrials any more
than we can about that of terrestrials?

The time appears to be fast approaching when we will
be forced to abandon the cultural fairytale most of us have
adopted by default: that we are alone or inaccessible in the
universe and that terrestrial humans can comfortably and
arrogantly place themselves at the top of the pecking order of
living beings.

The Roper Report

We know how difficult it is to openly admit to and
acknowledge UFO sightings. It is even more difficult to
accept the so-called abduction phenomenon, in which otherwise normal people claim to have been taken on board a
craft, examined medically and asked, or more frequently
compelled, to participate in various activities, including apparent efforts to breed a human-alien hybrid race.

Yet three carefully crafted polls taken by the mainstream Roper Organization show that at least two per cent of

the adult American population claim to have been abducted by aliens! This conclusion is based on questionnaires sent to nearly 6000 respondents. Eighteen percent reported "waking up paralyzed with a strange person or presence or something else in the room." Interestingly, the incidence of these unusual phenomena appears to be higher for "influential" Americans—those who are socially or politically active and have higher incomes. Twenty-eight per cent of this population reported having woken up with the sense of a strange figure present.

It has been argued that the shocking numbers, apparently totalling millions in the United States alone, may actually be low. This is because abductees seem in most cases to have been put into a state of induced amnesia during these experiences and tend initially to recall them as if they were nothing more than vaguely-remembered dreams. But often they come fully to memory under hypnosis, and in some cases are recalled spontaneously in complete detail.

What do we make of all this? John E. Mack, a Harvard professor of psychiatry, has used the Roper Poll results to advise the mental health community to be aware of the pervasiveness of the phenomenon and not to automatically dismiss abduction reports as being delusional. Mack's own story in many ways parallels that of the doctor portrayed by Richard Crenna in the 1992 CBS-TV miniseries *Intruders*, whose title was taken from the best-selling book of the same name by Budd Hopkins. The series, which was based on hundreds of real-life accounts, explored the moral challenges that confront a mainstream psychiatrist who discovers that his clients have been experiencing alien abductions.

Some have asked whether virtually the entire human race has had these things happen, whether we may be as animals in a vast cosmic zoo, and so forth. Others point out that whether or not such "silly" or "unthinkable" scenarios are literally true, entertaining them can be an extremely useful exercise in broadening one's perspective. After all, any mean-

ingful extension of our horizons virtually demands that we be
open to a radically expanded view of things. Today's heresy is
often tomorrow's truth.

"UFO research," said the late J. Allen Hynek, "is
leading us kicking and screaming into the science of the twen-
ty-first century." Many experts consider Hynek to have been
the father of UFO research, and his comments appear to be
prophetic. In terms of funding control and peer pressure,
those two ever-present forces that keep most of us well within the box, there is no better example of second-class citi-
zenship than one who has the audacity to inquire into
UFO-related matters. Denial and dismissal of the reality of
the UFO phenomenon are still rampant within governmental circles, the mainstream scientific community and even
some new-science organizations. I sometimes call ufology
the orphan child of the black sheep of the sciences.

    And yet, for those who will examine it, the evidence is
overwhelming that UFOs and their occupants are quite real
and are pressing ever harder against the walls of disbelief
erected and maintained by our traditional institutions and
authorities, as well as by our inside-the-box consciousness.
As reports come in from all continents, and as our own deter-
mination to know the truth grows, the walls are rapidly coming down.

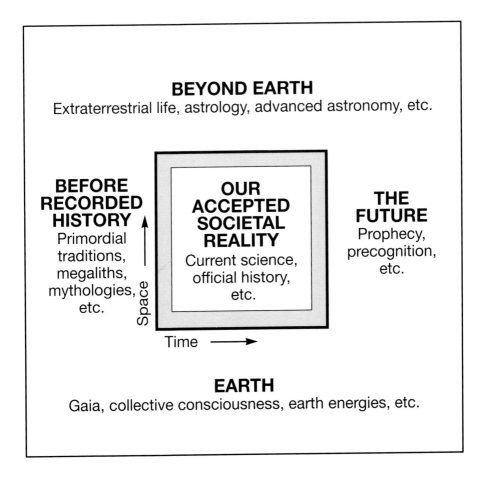

Fig. 1. Author's rendition of John Rossner's "collective box metaphor" diagram (See Chapter 2).

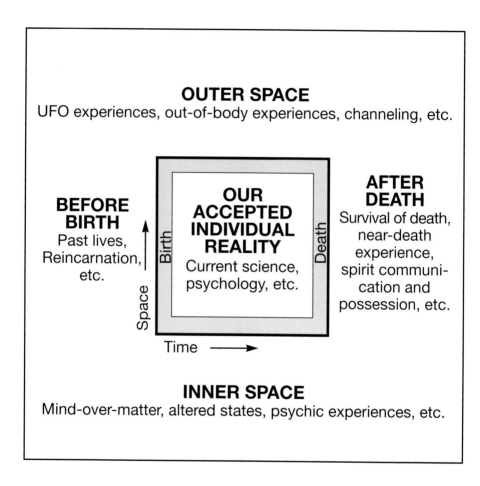

Fig. 2. Author's "individual box metaphor" diagram (See Chapter 2).

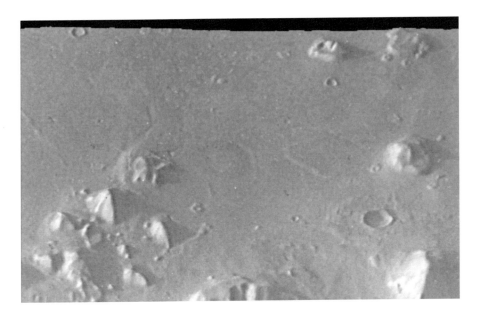

Fig. 3. The Face on Mars (upper right) and nearby anomalous features (See Chapter 2). Nearly a dozen possibly artificial structures have been discovered in NASA's 1976 Viking photos by several teams of independent investigators. After extensive analysis of these images, some researchers have argued that sophisticated mathematical relationships and alignments exist among these objects. Having dismissed these claims without investigating them, scientists at NASA and elsewhere maintain that the Face and other enigmatic objects are simply "tricks of light and shadow" or the products of random erosion. Digital perspective transformation of NASA Viking photo courtesy of Dr. Mark J. Carlotto.

Figs. 4(left) and 5(above). Bob Jahn's "pinball machine" ("Murphy") after the run performed by my son and I in which we willed the balls to go to the right (See Chapter 3). Note that the center bin contains fewer balls than the adjacent bins and that many more balls fell into right bins than into their corresponding left bins. Fig. 4 photo by the author, Fig. 5 photo courtesy of the PEAR labs.

Fig. 6. Bob Jahn and Brenda Dunne pose in front of their random event generator. This device has been employed in psychokinesis experiments whose statistically significant results have shocked the scientific world. Photo by the author.

Fig. 7. In this photograph a group of visitors witness Thomaz Green Morton materializing perfume (See Chapter 4). It is issuing from the back of his hand and dripping down his fingers to my outstretched palm.

Fig. 8. Moments later, here I am participating in this impressive
materialization.

Fig. 9 In this view, Thomaz is about to split a fork in two using his psychokinetic ability.

Fig. 10. A few seconds later the prongs of the fork (not to be confused with my jacket collar in this photo), along with the hot dog, have dropped to the table.

Fig. 11. Here we are posing with the broken fork. I hardly look the part of the objective scientist-observer, but Thomaz' paranormal feats are so obvious and repeatable as to make such posturing seem quite unnecessary.

Fig. 12. During my November, 1990 visit to his laboratory, the late Marcel Vogel poses next to his mass spectrometer (See Chapter 4). The bottles in the background contain wine which Vogel fermented instantaneously using one of his fluid restructuring units. Photo by the author.

Fig. 13. Walber Pinto, a Brazilian student of Marcel Vogel's, is using a Vogel crystal to heal me of jet lag after the long flight to Brazil.

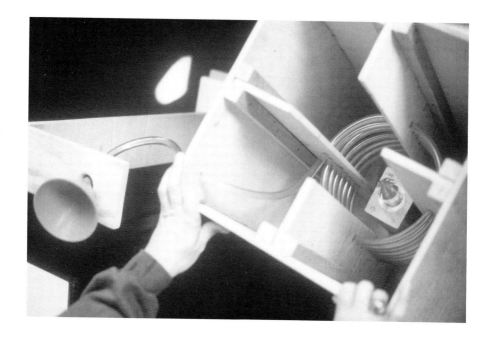

Fig. 14 Top view of a Vogel fluid restructuring unit. A liquid is poured into the funnel, spins through the energy field of a Vogel crystal, and emerges in treated form. Photo by the author

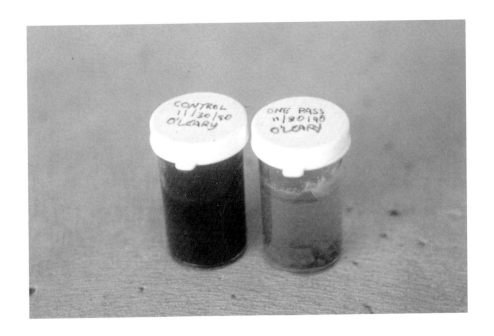

Fig. 15. This photograph shows two samples of apple juice injected with mold six months earlier. The control sample, left, is thoroughly contaminated with the mold, whereas the sample on the right, which has passed once through a Vogel fluid restructuring unit, remains clear; the mold has not spread. Photo by the author.

Fig. 16. New Scientist Tom Bearden (See Chapter 6) during our intense weekend session in May, 1991. Photo by the author

Fig. 17. Willis Harman, president of the Institute of Noetic Sciences (a true pioneer of New Science) and I are enjoying the view after a climb to a Blue Ridge summit in Virginia.

Fig. 18. Here I am (lower right) preparing myself for an encounter with Sai Baba (See Chapter 8). Photo by Gary Conrad.

Fig. 19. Sai Baba appears for his morning darshan at his Whitefield ashram near Bangalore, India. Photo by the author.

Fig. 20. With a flick of is hands, Sai Baba produces *vibuti*, the holy ash.
Photo by the author.

Fig. 21. When our group was granted a private audience with Sai Baba, he produced for me, by a wave of his hand, this beautiful ring. Photo by Kevin Stevens.

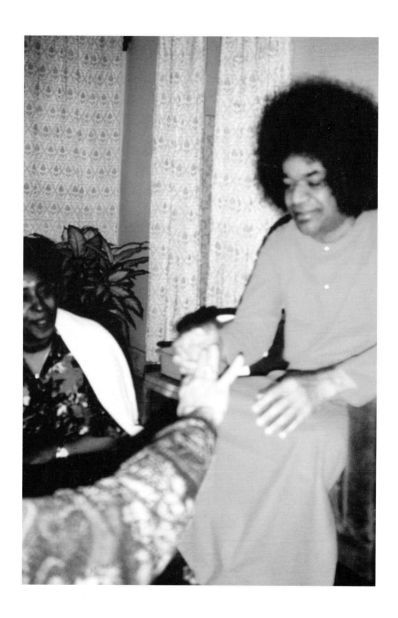

Fig. 22. Here Sai Baba is checking the fit of my ring, which was at first a bit loose and which he corrected perfectly simply by blowing on it. Photo by Kevin Stevens.

Fig. 23. Sai Baba is only 5 feet tall and is 65 years old. Here I am feeling quite ecstatic about our encounter. Photo by Kevin Stevens

Fig. 24. Sai Baba gives my ring one more glance. Looking on is B.N. Srikantaiya, a Sai Baba devotee who helped us gain our audience. Photo by Rick Butterfass.

Fig. 25. Devotees await Sai Baba's appearance as he strolls toward them, still holding my book. Photo by Kevin Stevens.

Fig. 26 Our group of American crop-circle tourist-researchers.

Fig. 27. English crop circle investgator Richard Andrews is dowsing the strong energy field associated with a recent crop formation. Photo by the author.

Fig. 28. My dowsing rod spun like the rotor of a helicopter near the center of one of the English crop formations.

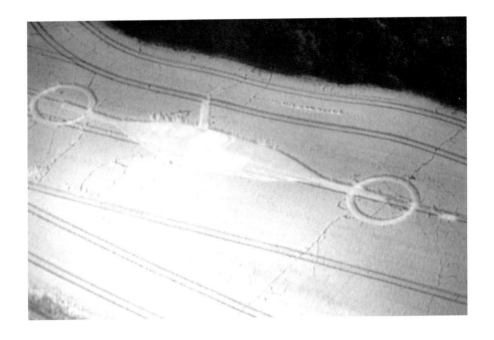

Fig. 29. Thirty-two of us from the United States walked into this mysterious crop formation in southern England the day after it was created (See Chapter 9). The energy felt by this group was extraordinary; some laughed, others cried, prayed and meditated. Photo by Kevin Stevens

Fig. 30. Participants discussing research proposals at the 1992 International Association for New Science UFO thinktank/retreat in Colorado (See Chapter 10). Photo by the author.

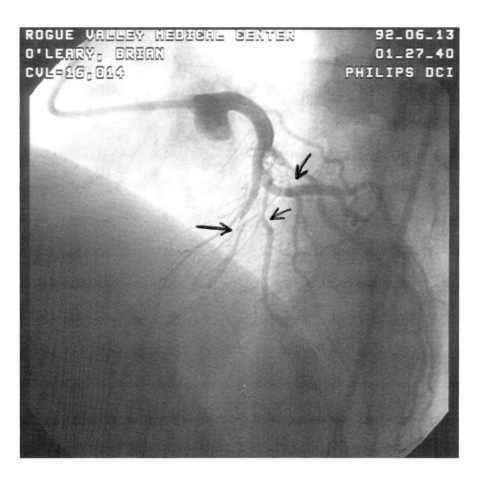

Fig. 31. In this angiogram of arteries leading into my heart, the left anterior descending branch was blocked (left arrow), causing my May 27, 1992 heart attack (See Chapter 11). Arrows to the right indicate moderate blockage of some tributary arteries. Photo courtesy of the Rogue Valley Medical Center and Dr. Brian Gross.

Fig. 32. In this photograph of the July, 1991 solar eclipse in Kona, Hawaii, a hole has opened up in the dense, obscuring clouds after three of us focused our energies on "vaporizing" them (See Chapter 11). After the four critical minutes of totality the clouds closed in once again. Photo by the author.

11

Unveiling the
Next Millennium

Del Mar, California, May 26, 1992

Having spent the previous five days at the UFO retreat and symposium in Colorado, I was looking forward to several days' rest and relaxation with my partner Meredith Miller, a talented Southern California artist. On the first day we passed the time in her room, which commanded a beautful view of the Pacific Ocean. It was a welcome relief after a stretch of hard but rewarding work.

It was late in the evening now. We had spent a wonderful day together and I felt a special flow of energy within me. As I was about to drift off to sleep I became aware of some nausea, dizziness and perhaps some digestive disturbance, so I rose and went to the bathroom.

As I was about to return, my dizziness got the better of me. I blacked out and fell to the floor, knocking myself out cold and receiving a cut and bruise on my forehead. As Meredith heard the thump of my fall she ran into the bathroom and saw me coming back into consciousness. I was "out" for only a few seconds, but was stunned and disoriented.

My recollection of those brief seconds was dreamlike and timeless. I recall having felt a flash of red light, then white light accompanied by loud rushing sounds. It felt as if I were being zapped by lightning, yet I experienced a comforting warmth as if I were swimming in some kind of golden nectar. This powerful, otherworldly interlude seemed to have been filled with a multitude of events that somehow were compressed into a tiny interval of earthly time and that quickly faded from memory. In some ways the incident reminded me of the near-death experience I had had in 1982 during an auto accident, which I described in my previous book, *Exploring Inner and Outer Space.*

As I slowly came to, I began to feel severe chest pains, continuing dizziness, nausea and a cold sweat. In a short while, assisted by Meredith, I crawled back into bed. For the next two hours I felt the most intense chest pains I had ever experienced, yet something told me that it would be best for me not to be taken to a doctor's office or hospital emergency room; that I would be safer resting in Meredith's loving care, away from the antiseptic medical environment.

Eventually the pain subsided. The nausea and dizziness stayed with me for another two to four days, while I fasted on sparkling water and a little fruit. Oddly enough, during this time a major construction project began in the road next to the house. Jackhammers, the beeping of trucks backing up and the yelling of construction workers were to carry many of the next days, yet I felt secure here, and quickly recovered my strength.

I recalled during this time that changes in consciousness and the emergence of extraordinary abilities have often been associated with traumatic experiences. In the case of Thomaz Green Morton it was a lightning strike, in that of Uri Geller severe shock, and with Sai Baba it was a scorpion bite. These consoling facts raced through my mind as I contemplated the fearful possiblity that this had been a heart attack, plain and simple, and that I might die prematurely just as my public and private lives were entering a reward cycle.

My fear was justified by the fact that my father had died suddenly and unexpectedly of a heart attack at the age of 67 and that his brothers and father also died of heart attacks. It is clear that I may have have inherited this unfortunate genetic predisposition.

My cholesterol and blood pressure were basically OK, although occasionally pushing the upper limits of the normal range during stressful periods. I have tended to overindulge at times, and over the years have experienced a fair amount of heartbreak both in personal relationships and in professional affiliations. At this time I had also been maintaining a punishingly stressful travel and lecture schedule.

So one week later, as soon as I arrived back home in Oregon, I went to a local doctor who took a cardiogram. It looked fine, and I celebrated the fact that it appeared I hadn't had a heart attack. It had, instead, perhaps been a case of food backing up into my esophagus, causing the chest pains and nausea. The doctor then asked me to return the following week to take a treadmill cardiogram—a definitive check to rule out the possibility that this had been a heart attack. Meanwhile, I resumed my normal life feeling relieved that I was all right.

On Thursday, June 11, I arrived at the doctor's office in my running clothes eager to have at the treadmill; I recalled my astronaut medical exam, in which I had excelled in such tests. They first took my rest cardiogram, and it looked strange to them because it didn't resemble the one taken the previous week. They then called in an internist, who agreed that this was not normal.

The doctor then faxed the mismatching cardiograms to a cardiologist in nearby Medford along with the results of a blood test showing that I also had anemia. I had less than two-thirds of the amount of blood I should have had, and the blood cells were much too small. Serious business.

The next day the cardiologist took a series of cardiograms that showed my heart beating irregularly, a blood clot on an interior wall of the heart, and the left anterior descending (LAD) branch artery completely clogged.

I had indeed had a heart attack, one which had put

me at a one-in-ten, or higher, risk of dying. I had just played a round of Russian roulette which fortunately I won. Would there be future rounds? I did not know.

The doctor immediately ordered me to the hospital the next day for an angiogram of the arteries leading into the heart. This is a process in which a catheter, a long, slim, hollow tube, is inserted in the groin area and pushed up through the arteries into the heart. Then a special colored liquid is injected that can be seen in x-ray films.

The results confirmed the doctor's earlier suspicions that the LAD artery was clogged (Figure 31). Some branches were also partially clogged, but there was only a trace of fat built up in the other main arteries. I didn't need to have bypass surgery, thank God.

Signs and Omens?

During those two tumultuous days of diagnostic procedures, it was unseasonably cold in southern Oregon at a time when it should have been sunny and hot. The sky showed an ominous overcast, as if to remind us of the gloom of the situation. But the day before I entered the hospital, and again when I was discharged, reassuring rainbows appeared in the sky. A similar symmetry of rainbows happened here as I was heading off and coming back from my trip to India. I had never seen any other rainbows in the nearly one year I lived here.

I am beginning to perceive events in nature and timely communications from friends as symbols reflecting a deeper reality. What could be more dramatic than black clouds and rainbows bracketing these important happenings? These seeming synchronistic events were amplified by a beautiful letter awaiting my June, 1992 return to my home in Oregon by B. N. Srikantaiya, the devotee of Sai Baba who had facilitated our meeting with the swami. In the letter he pointed out that all life emanates from the heart, a word that can be divided into the words "he" and "art." This astonished me because I had had, for the first time in my life, the very same thought earlier that day on the airplane as we flew into Oregon, and I had shared it with Meredith. The thought was compelling and had a deep, haunting quality to it, yet I felt that its full meaning would not be illuminated until later. Patience, and a willingness for things to unfold, is part of the process of what we have called the New Science.

Synchronistic events are not the only "omens" worth noting; often the insights and predictions of psychics, whether taken literally or symbolically, can provide meaningful clues to subtle events that may be hidden behind the scenes of ordinary life. For example, a woman with a reputation as a psychic and clairvoyant lived downstairs from Meredith and was asleep at the time I experienced my heart attack. In her dreams that night, she saw me being medically examined by a group of hostile aliens who were sticking tubes and needles into me. Although this may be stretching things a bit, I have wondered whether this woman's dream might not have had an element of truth to it, given the peculiar time-distortions and other psychic experiences that accompanied the onset of my physical symptoms. Could my heart attack have been triggered by a stressful or traumatic momentary encounter with extramundane beings rather than being nothing more than the result of a gradual buildup of plaque in my arteries?

New Science would not forbid asking such radical questions but rather would test them with a combination of retrospective cardiology, hypnotic regression and analyses of similar cases.

Interestingly, my cardiologist was somewhat puzzled by the flashes of light and gushes of sound I experienced prior to feeling the chest pains; these are by no means typical precursors to the physical onset of a heart attack. Perhaps, he theorized, my heart may have stopped or gone into spasm momentarily for unknown reasons, allowing the triggering clots to form. My detective work on these questions has only begun.

Back to the Physical Heart

As I write this I am on medication to thin out my blood so it doesn't clot again and cause future heart attacks. I also have a nitroglycerine spray to expand my arteries if I should again feel chest pains. I hope and trust I'll never have to use it. I am also on iron pills, which have already raised my blood hemoglobin.

Could the existing clots cause more havoc? Possibly. Theoretically, part of the clot that formed in the inner walls of the heart could dislodge itself and "go North," causing a stroke that could take my life within seconds. How much of the heart muscle died as a result of the attack? The doctor suggested 15 per cent.

Mysteries remain that need to be unscrambled: What actually did cause the heart attack? How did I lose all that blood and contract anemia? (My previous blood test seven months earlier had shown none whatsoever.) In spite of my family's experiences and the stresses of my lifestyle I had had no personal history at all of either condition. Was this a purely physical cause-and-effect situation or could the attack have been triggered by something anomalous? I was certainly prepared to entertain a range of possiblities.

While all this is too early for 2020 hindsight, my heart attack has clearly been a wake-up call. Not only do I need to eat well, slow down, exercise regularly and minimize stress, I need to heal myself of this condition and prevent future occurrences. Since the heart attack I have lost over 15 pounds and am in the top quarter for my age group for physical endurance in a treadmill test.

Positive thinking, visualizations and so forth are a tall order for conventional medical science, but a routine consideration in the context of New Science. As I described in

Exploring Inner and Outer Space, I once used mind-over-matter techniques to heal my own "incurable" knee condition. There is no reason why I should not do the same with my heart condition and anemia.

Numerous encouraging examples of dramatic self-healing are revealed in the literature, such as the late Norman Cousins having laughed himself to health, *Dr. Dean Ornish's Program for Reversing Heart Disease*, and the Spindrift experiments on the effectiveness of prayer in healing heart patients, as documented by Larry Dossey in *Recovering the Soul*.

My work on my new life has barely begun, but I welcome this chance to be kinder to myself and to grow more consciously into a stronger, healthier, more balanced human being.

To paraphrase an old aphorism, "Metaphysician, heal thyself!"

Kona, Hawaii, July 11, 1991

We now flash back to the place where I envisioned myself to be in the year 2020. Today was to be the day of the total solar eclipse that would turn the Big Island into night for about four minutes. Because the eclipse was to be at seven in the morning during the summer, the probability for clear skies on the Kona side of the island was very high—about 95 per cent. Unfortunately, it began to look as if we would hit the one in twenty.

We awoke that morning at four o'clock in an unexpectedly drizzly giant-fern forest to set out along the high road toward the dry Kona side of the saddle between Mauna Kea and Mauna Loa. At daybreak the clouds remained disappointingly thick, with only an occasionally veiled clear spot passing by momentarily to reveal the Sun being chewed up by the Moon.

I was with my friends Roger and Lydia Weiss, who are no strangers to New Science. They were board directors of the New Thought Church of Kona, espousing a spiritual philosophy that strongly parallels some of the basic tenets of New Science.

Recalling Richard Bach's novel *Illusions*, in which the "reluctant messiah" Donald Shimoda was able to vaporize clouds, we decided to attempt the same. We started by doing the same kind of aikido breathing exercises that had allowed us to bend spoons. We prayed and intensely focused our energy on the thick clouds blocking the early morning sun. Then, just as totality began, the clouds began to part in precisely the right spot. I took some pictures of the blocked sun and its corona (Figure 32) through the remaining thin veil of clouds. Then, as totality was about to end, the thick clouds returned.

We later found out that, apart from the professional astronomers on top of Mauna Kea (which was off-limits to the rest of us), only a handful of people on the Big Island managed to see any of the eclipse. Was this just the luck of the Irish? Perhaps so, but the timing was uncanny.

Interestingly, the same thing happened during the only other solar eclipse I had observed. This one was in Maine in 1962, when I was an astronomy graduate student at Georgetown University. The fact that some priests were there may have helped. Luck of the Irish, for sure!

Kona, Hawaii, January 27, 2020

Resuming the story that began this book, I walked in front of an audience of over 1000 people, looked at my notes for a few moments, then put them down.

"I have been here in Kona during many special moments in my life," I began. "But this one is most special. It feels like coming home into my heart."

"I've been here over the years during some very important milestones in my life: in 1970 as a mainstream planetary scientist; in the early 1990s as a budding new scientist, cloud vaporizer, heart patient, nature appreciator, lecturer, workshop leader, dolphin seeker, lover, and volcano watcher. I love smelling the salt air and flowers and hearing the surf roar."

I looked out toward the sea and took a deep breath.

"Today I was asked to share my views of the history of the New Science movement in a one hour speech, but I feel compelled instead to reiterate and review its most important principles—ones that sometimes slip through the sieves of time."

"During the early 1990s we were fumbling and groping with these principles in the face of adversity from our peers and our funding sources. Some of us felt the loneliness of a Galileo in the presence of the Catholic Church or of a Founding Father in the face of the English King. We felt somehow driven by radical ideals most people would not have dared to think were possible, or able to be implemented."

"Yet, though our work was unsupported, something kept us going. We were pretty crazy, I suppose. Many of us were unqualified for mortgages, wondered how the next month's rent would be paid, and lived as if there would be no tomorrow. Some of us wondered whether there'd be any tomorrow."

I reached for a glass of sparkling water, took a sip, looked over the audience, and continued.

"The first principle of New Science is the willingness to venture outside the box. Rewarding orthodoxy and discouraging innovative inquiry in the face of clear evidence is no longer thinkable."

"So we have now built bigger boxes. But we must never forget that bigger boxes are still boxes. *Keep looking outside!* That's where you'll find most of the action. That's where you'll find cream to skim. Remember that the first priority of science is to explore the unknown. Rigor must follow, of course. . . but it must never be allowed to lead!"

I paused, looking out at the hibiscus and palm trees, and somewhow began to feel mesmerized by the natural surroundings. All of a sudden I felt compelled to go out into it rather than present a long speech.

"The second principle is that we always need to consider reality in terms of the interconnectedness of all things. Reductionism is fine as far as it goes, but it doesn't go very far." I heard some knowing snickers in the audience. "Look at how far we got once we balanced holistic health with conventional medicine and how fast things unfolded in environmental science once we opened ourselves to acknowledging many points of view.

"Thank God those dark days of academic and bureacratic calcification, hyper-specialization and insanely diminishing returns are behind us forever. We all atrophied in our own boxes and practially lost our reason for being!

"We now recognize the necessity of combining the disciplines of art, music, the humanities and the sciences into an interactive whole. We now know how much enjoyment there is in sharing and pooling our talents and discoveries.

"A third principle is the need to discard the assumption that we are all like machines driven by their parts, rather than participants in an intelligent, overarching scheme of things.

"Gone are the days of our denial of a higher order in nature. Gone is the illusion that most of the action in the universe takes place on the physical plane. We now know we have to revise our science to open up new possibilities of model-building within higher realms of being, beyond mass, energy, space and time.

"The fourth and last principle involves the ethical context within which the entire enterprise is undertaken. Greed, manipulation and the hoarding of wealth can no longer carry the day. We nearly destroyed ourselves and our planet doing that. Our extraterrestrial teachers have now begun to reveal to us the magnificent ethical principles recognized by those who operate under cosmic law.

"As a species, we now know how serious was our avoidance of the understanding of cause and effect—both in the realm of mundane, physical interactions and in the nonphysical dimensions of karma.

"In mistreating our animals, plants, the Earth and ourselves, we invited scavenging alien activity which was traumatic but really nothing compared to what we had done all by ourselves. Fortunately, we can now track all this from a more cosmic perspective.

I once again looked toward the surf and breathed in the fresh air.

"It's hard to believe we were in such darkness as recently as the 1990s. May it never happen again. May we love and enjoy ourselves from this position of 2020 hindsight, a lesson in the transformation of our being to the incredible multitude of awarenesses and joys we now feel for being human on this planet at this time.

"And it's encouraging to us older folks that we now know we have an eternal existence to look forward to, either inside or outside these Earthship spacesuits—our bodies. No small matter.

"Well, this speech is about over. I've gone on long enough. I yield the rest of my time to wading barefoot in the surf. If anybody would like to join me, take your shoes off and follow me down to the beach!"

As I began to dig my feet into the sand and felt the water lap over my ankles, I felt the warm rays of the sun mingle with a strange golden light that seemed to be emanating from my feet. It began to move up my body, bathing me like the golden nectar I had felt back in 1992.

Toward the land I saw a brilliant rainbow, the second of a pair I had seen today. Pairs of rainbows, golden nectar...

Then my whole body seemed to glow. My heart filled with the nectar and light, and suddenly I felt weightless. The whole notion of physical age seemed almost laughable; an illusion of my earthly existence that I had brought with me right up to this very moment, one that was derived from the same set of illusions that drove the old science.

In that moment I knew, because I remembered, that my essence would carry on in a never-ending exploration of a dazzling universe among dazzling universes.

I finally released the last of my fears and became the light I really am.

References

Chapter 1:

Kuhn, Thomas, *The Structure of Scientific Revolutions*, University of Chicago Press, Chicago, IL. (1970).

Chapter 2:

Campbell, Joseph, interviewed by Eugene Kennedy, "Earthrise: The Dawning of a New Spiritual Awareness", *The New York Times Magazine*, April 15, 1979, pp.14-15.

Carlotto, Mark J., *The Martian Enigmas: A Closer Look*, North Atlantic Books, Berkeley, CA. (1992).

O'Leary, Brian, *Exploring Inner and Outer Space*, North Atlantic Books, Berkeley, CA. (1989).

O'Leary, Brian, "Analysis of Images of the Face on Mars and Possible Intelligent Origin," *Journal of the British Interplanetary Society*, Vol. 43, pp. 203-208 (1990).

O'Leary, Brian, "New Science Potential," *New Science '90*, edited by Maurice L. Albertson, Rocky Mountain Research Institute, Room 203, Weber Building, Colorado State University, Fort Collins, CO 80523 (1990).

Chapter 3:

Fishman, Steve, "Questions for the Cosmos," *The New York Times Magazine*, November 26, 1989, pp.50-55; also "The Dean of Psi," *Omni* Magazine, September 1990, pp. 42-46 and 88-90.

Jahn, Robert and Brenda Dunne, *Margins of Reality*, Harcourt Brace Jovanovich, San Diego, CA. (1987). Technical papers can be ordered through the PEAR Labs, School of Engineering and Applied Research, Princeton University, Princeton, NJ 08544-5263.

O'Leary, Brian, "The Presence of Ice in the Venus Atmosphere as Inferred from a Halo Effect," *Astrophysical Journal*, Vol. 146, pp. 754-766 (1966). This was my first suggestive evidence for a Venus halo effect.

O'Leary, Brian, "Venus Halo: Photometric Evidence for Ice in the Venus Clouds", *Icarus*, Vol. 13, pp. 292-298 (1970). This was stronger, further evidence of the Venus halo effect.

O'Neill, Gerard K., *The High Frontier*, Wm. Morrow, New York (1977).

Ward, Dennis and Brian O'Leary, "Search for a Venus Halo Effect during 1970," *Icarus*, vol. 16, pp. 314-317 (1972). This was negative evidence for the effect.

Chapter 4:

Pulos, Lee and Gary Richman, *Miracles and Other Realities*, Omega Press, San Francisco (1990). This is a biography of Thomaz Green Morton.

Marcel Vogel's work is documented in the *Psychic Research Incorporated newsletters*. More information can be obtained from Vogel's former assistant Jenet Grover, Box 2870 MCCA, Estes Park, CO 80517.

Chapter 5:

Hawking, Stephen, *A Brief History of Time*, Bantam Books, New York (1988).

Jahn, Robert, cited in Chapter 2.

Stone, Robert B., *The Secret Life of Your Cells*, Whitford Press, West Chester, PA. (1989). This book is a good summary of Cleve Backster's work.

Talbot, Michael, *The Holographic Universe*, HarperCollins, New York (1991).

White, John, *The Meeting of Science and Spirit*, Paragon House, New York (1990).

Zukav, Gary, *The Seat of the Soul*, Simon & Schuster, New York (1989); also his presentation "Science snd Spirit," *New Science '90*, edited by Maurice L. Albertson, Room 203, Weber Building, Colorado State University, Fort Collins, CO 80523 (1991).

Chapter 6:

Bearden, Thomas E., *The Excalibur Briefing*, Tesla Book Company, Greenville, Texas and Strawberry Hill Press, San Francisco (1988 and 1990); *AIDS Biological Warfare*, Tesla Book Company, Greenville, Texas (1988). Other Bearden papers and books may be ordered from the Tesla Book Company or through Bearden at 2311 Big Cove Road, Huntsville, AL 35801.

Bohm, David, *Wholeness and the Implicate Order*, Routledge & Kegan Paul, London (1980).

King, Moray B., *Tapping the Zero-Point Energy*, Paraclete Publishing, P.O. Box 859, Provo, UT 84603 (1989).

Chapter 7:

Pleass, C.M. and N. Dean Dey, "Conditions that Appear to Favor Extrasensory Interactions between Homo Sapiens and

Microbes," *Journal of Scientific Exploration*, Vol. 4, pp. 213-231 (1990).

Radin, Dean I. and George R. Cross, "Sequential Ordering Effects in a Perceptual Experiment: Testing the Nature of Prayer and Psi," GTE Laboratories - Chantilly, 15000 Conference Center Drive, Chantilly, VA 22021-3808 (1991).

Rockwell, Theodore, "Reactionary Revolutionaries: Is the Shoe Now on the Other Foot?" (1991).

Spindrift's Occasional Newsletter, P.O. Box 5134, Salem, OR 97304-5134: The VIUR test is written up in some of the 1991 newsletters.

Stone, Robert B., cited in Chapter 5.

Chapter 8:

Heraldsson, Erlendur, *'Miracles are my Visiting Cards'*, Century, London (1987).

Chapter 9:

The Crop Circle Enigma, edited by Ralph Noyes, Gateway Books, Bath, England (1990) and revised in 1991.

Delgado, Pat and Colin Andrews, *Circular Evidence*, Bloomsbury Publishing, London (1989) revised1990.

Chapter 10:

Greer, Steven M., Director's Message, *The CSETI Newsletter*, P.O. Box 15401, Asheville, NC 28813 (April 1992). This describes the Belgian and Gulf Breeze close encounters of the fifth kind.

Hopkins, Budd, *Intruders*, Random House, New York (1987).

International Symposium on UFO Research, proceedings edited by Maury Albertson and Margaret Shaw, The International Association for New Science, 1304 S. College Ave., Fort Collins, CO 80524 (1992).

Unusual Personal Experiences, an analysis of the data from three national surveys conducted by the Roper Organization, Bigelow Holding Organization, 4640 South Eastern, Las Vegas, NV 89119 (1992).

Chapter 11:

Bach, Richard, *Illusions*, Delacorte Press/Eleanor Friede, New York (1977).

Cousins, Norman, *Anatomy of an Illness*, Norton, New York (1979).

Cousins, Norman, *Human Options*, Norton, New York (1981).

Dossey, Larry, *Recovering the Soul*, Bantam Books, New York (1989); also, the Spindrift work on the effects of prayer on heart patients can be obtained from the address listed in Chapter 7.

Ornish, Dr. Dean, *Dr. Dean Ornish's Program for Reversing Heart Disease*, Ballantine Books, New York (1991).

Appendix:

Some New Science Organizations

Many organizations have been established in recent years to serve the growing needs of the New Science community. I shall briefly mention, and share my personal impressions of, four of the larger organizations in the United States having annual meetings that typically attract over a hundred attendees. These four groups are the Society for Scientific Exploration, the United States Psychotronic Association, the Institute of Noetic Sciences and the International Association for New Science.

The Society for Scientific Exploration

The SSE is a group of elite scientists. Membership is by election; a Ph.D. and a research track record are normally required. The organization meets once a year, usually on the East Coast during May and June.

I had the pleasure of attending the SSE tenth annual meeting held in Charlottesville, Virginia on May 23-25, 1991. At this meeting papers were presented by Robert Jahn and his group at Princeton on their positive results on psychokinesis, by Ian Stevenson and his group from the University of Virginia on reincarnation and maternal-impressions research, and by Fritz Popp, Michael Pleass and others on recent results in biocommunication. Several excellent contributions were also made on subjects ranging from new physics to near-death experience to UFOs.

While the SSE emphasizes open inquiry into areas it considers to be anomalous, it seems also to be open to the possibliity that new scientific paradigms may embrace these areas, rendering them non-paranormal. Such flexibility on the part of mainstream scientists affiliated with prestigious universities is most encouraging to many of us in the New Science movement.

Information about the SSE can be obtained from L.W. Fredrick, Secretary, SSE, Box 3818, Charlottesville, VA 22903.

The United States Psychotronic Association

Another meeting I attended during 1991 was the 17th annual conference of the United States Psychotronic Association, held between July 17 and 21 in Dayton Ohio.

Dedicated to the late Marcel Vogel, who had been active in the USPA, this meeting brought together several outstanding scientists in those New Science areas involving the measurement of free energy, as well as subtle energies, mind-body interactions and their effects on healing and on the restructuring of fluids. Papers were presented by Moray King on free energy, by Cleve Backster on biocommunications, by Glen Rein on quantum biology, by Beverly Rubik on biogravity, by Sarah Hieronymous on homeopathy, and by Ted Rockwell on scientific qigong.

The quality of discussion at the USPA is high, as befits an organization that has been doing New Science for almost two decades. Information on USPA membership and activities can be obtained from Bob Beutlich, president, 2141 Agatite, Chicago, IL 60625.

The Institute of Noetic Sciences

IONS was founded in 1973 by fellow astronaut Ed Mitchell. Willis Harman, President of IONS (Figure 16), has been of great help to us in establishing our own International Association for New Science (see below).

IONS has a large membership and is active in research and education. Its prime areas of focus include mind-body interactions, health and healing, emerging paradigms in science, society and business, creativity and human potential, conscious living/conscious dying, transpersonal psychology, personal growth, meditation and spiritual quest. IONS has steered clear of some of the more controversial areas of research such as UFOs, psychokinesis, precognition, near-death experience and reincarnation.

On June 26-29, 1992, the Institute of Noetic Sciences held its first international conference in Santa Clara, California. Presenters included Sam Keen, Willis Harman, Hazel Henderson, Rachel Naomi Remen, Charles Garfield, Ed Mitchell, Joan Borysenko, and Ken Pelletier.

For more information, contact IONS at 475 Gate Five Road, Suite 300, Sausalito, California 94965, (415) 331-5650.

The International Association for New Science

Founded by Maury Albertson and myself, this rapidly growing organization encourages open membership and has a research arm that is self-selecting.

Our first International Forum on New Science took place in Ft. Collins, Colorado, in September 1990. Invited keynote speakers and contributed papers are included in each annual conference held in September. The attendance at each of the first two conferences was about 400.

The IANS has also held a private think-tank/retreat and related public symposium on UFO and Extraterrestrial Studies, May 20-25, 1992 in the Denver, Colorado area. At the retreat, twenty-five leading researchers met to draft proposals for potential funding of UFO-related research (see Chapter 10 for details).

A specialty conference on free energy is planned for February 12-14, 1993, and on New Medicine during the first weekend of June, 1993. Information on the IANS can be obtained from Carol Singer, IANS Administrator, 1304 South College Avenue, Fort Collins, CO 80524, (303) 482-3731.

Index

Dr. O'Leary is available to give lectures and seminars on the subject of this book. For more information, write to him at 755 Tyler Creek Road., Ashland, OR 97520.